How to Enrich Geometry Using String Designs

Victoria Pohl
Professor of Mathematics
University of Southern Indiana

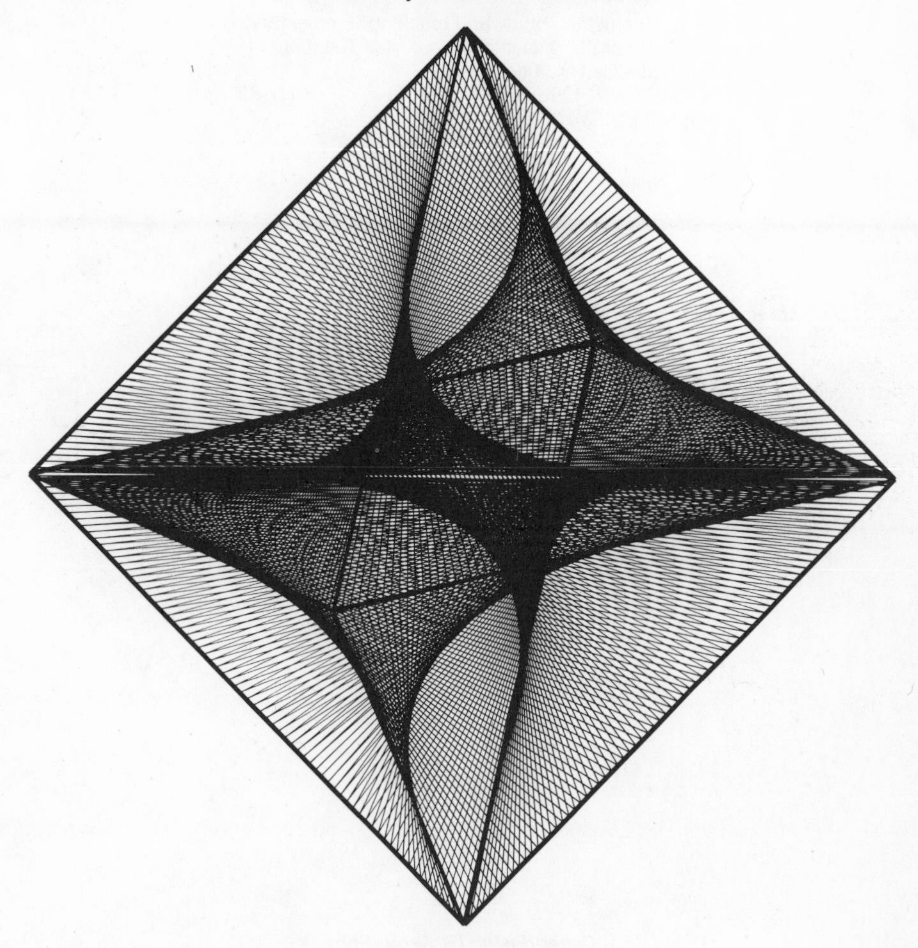

With drawings by the author

National Council of Teachers of Mathematics

Library of Congress Cataloging in Publication Data:

Pohl, Victoria.
How to enrich geometry using string designs.
Bibliography: p.
1. Geometry—Study and teaching (Elementary)
2. String craft. I. National Council of Teachers of Mathematics. II. Title.
QA462.P65 1986 372.7′3 86-5189
ISBN 0-87353-227-9

Cover design by Karen Schenk

Printed in the United States of America

Contents

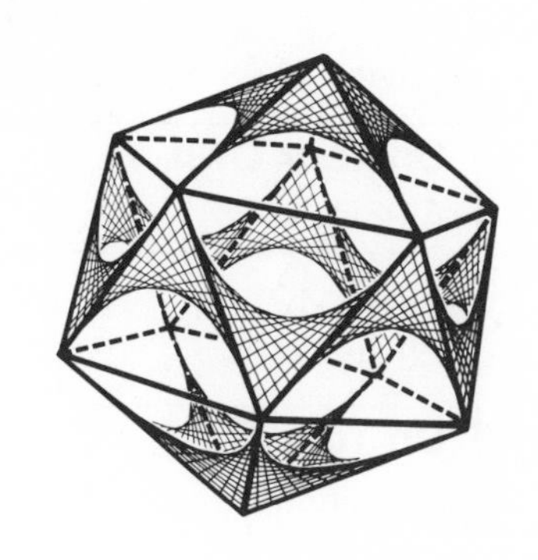

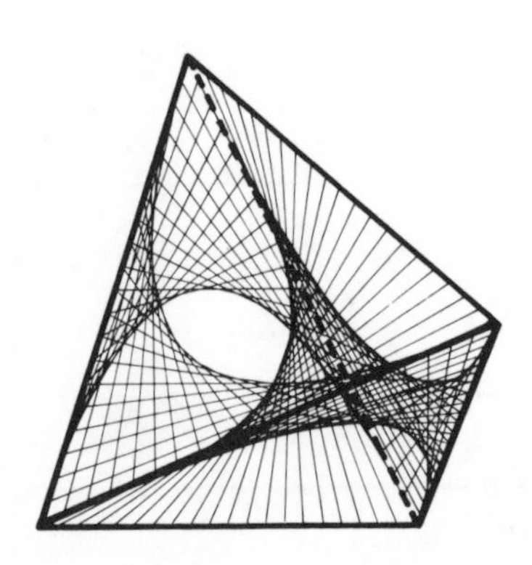

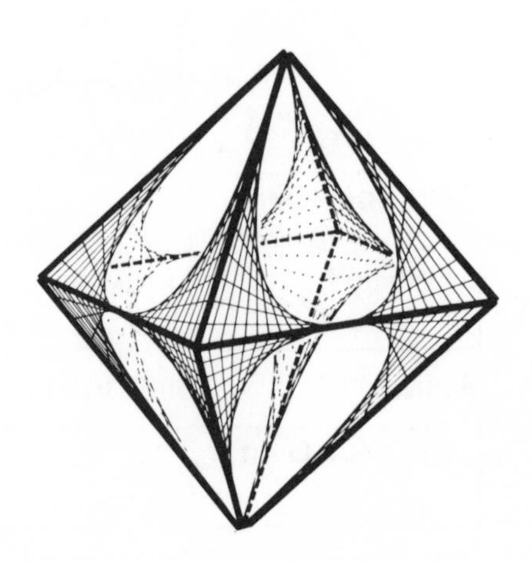

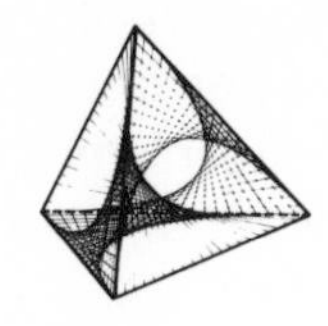

Introduction

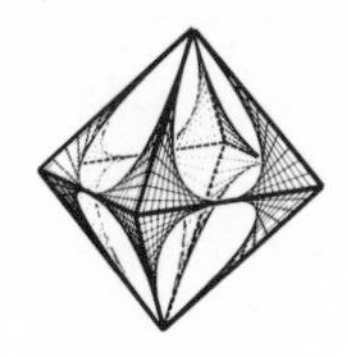

The activities in this book—constructing string designs on polygons and polyhedra—were created for grades 6–10, but their enjoyment is by no means restricted to these levels.

The designs with polygons, constructed on cardboard, are easier to make and take less time than those with polyhedra, which are constructed on plastic straws or wood dowels. Completed three-dimensional models, or even the two-dimensional ones, can be effectively displayed as mobiles. Slowly rotating polyhedra show off to advantage their attractive curves and surfaces.

Use the activities in your classroom as enrichment material for the entire class, a small group, or an individual. The punched and perforated pages can be easily removed and duplicated, then stored in a three-ring binder for further use. Numbered activities, which include step-by-step directions, are essentially arranged from easiest to most difficult. Those marked with an asterisk are more challenging than the others, and those with two asterisks, even more so. Many of these step-by-step activities are followed by related activities that encourage initiative and experimentation. These are designated by letters (A–N) and labeled "On Your Own."

"Related Exercises" can be used to aid class discussion on mathematical ideas or as homework. "Teacher Notes" indicate the appropriate grade level, the concepts being introduced, and skills required. (One of the most important requisite skills is not listed: the ability to follow directions.) They also provide answers for the exercises. Because of space considerations, these notes sometimes appear on the same sheet with the student materials rather than alone on the back of the sheet. In such cases you will want to cover that portion before duplicating the page.

These constructions offer abundant opportunity for students to learn mathematical vocabulary and concepts. Working with the designs on polyhedra also helps prepare high school students for the three-dimensional graphs and surfaces encountered in college calculus and analytic geometry. Although the activities are planned primarily for mathematics classes, they are also suitable for art (the models are beautiful) and reading classes. (Several field testers report using the activities as an exercise in following directions.)

Hints, Helps, and Hazards

Before you begin, you will want to know something about the strengths and weaknesses of various materials and how to avoid some of the potential hazards.

String. For each project, begin the weaving slowly until the pattern is familiar. For the straws, use thin cotton tatting or crochet thread, polyester sewing thread, or one strand of embroidery floss. For wood dowels or cardboard, these same types of thread or somewhat heavier cotton crochet thread works well. *Caution:* Polyester may sag later because it gradually stretches. Pull the strings tightly while building polyhedra with straws and while weaving, because sagging straws and limp strings detract from the beauty of a geometric model. When working with wood dowels or cardboard, wind the string on a spool, cardboard, or rolled paper and unwind it as you use it. For weaving on plastic straws, use no more than four feet of string on the needle; it tangles when longer pieces are used.

Colors. Coordinate the colors of string with the cardboard or straws. The use of contrasting colors makes an attractive figure. When using two or more colors of string on one model, choose colors with the same tone value, such as all bright colors or all pastels. For a more distinctive showing, string on a polyhedron should also differ from the color of the anticipated background—such as the walls of the room where it will hang if it is to be displayed as a mobile.

Glue. White clear-drying glues work well on the wood dowels. Rubber cement seems to work best for holding the plastic straws in place.

Cardboard. Cardboard should be stiff so it will not arch when the string is applied. A second layer of cardboard on the underside trimmed about one-fourth inch smaller all the way around strengthens a cardboard polygon. If colored cardboard is not available, paste colored paper on cereal box cardboard or corrugated cardboard. Make all markings on the underside of the cardboard unless otherwise specified.

Straws. Use a heavy needle and thin thread for dropping the threaded needle through the straws; otherwise, the needle and thread may get stuck inside the straw. Plastic drinking straws or stiff round stirring straws are suitable for the polyhedra. Slender straws make neater models. However, slender straws and long straws are more apt to bend. Punching the holes farther apart helps to prevent straws from arching but makes the model less delicate in appearance. If straws are bending, insert pieces of clothes hanger or dowels through them. A bit of rubber cement will keep these pieces from falling out.

Wood dowels. The sturdiest models are made with wood dowels. Making the notches takes more time, but the wood rarely arches during the weaving. The most attractive models have slender edges and notches close together. The careful person can do this on slender straws. The rest of us must be satisfied with reasonably impressive models made with short drinking straws—or we can choose wood dowels and make larger magnificent models like those shown in the photos on and inside both covers of this book.

Computer graphics. This book contains many designs that students can sketch on the screen of a computer. The drawing on the title page and the pictures of the models for Activities 14–18 and Designs 3–5 in Activity M were done by William McWorter on a personal computer. A sample program is included in the related exercises for Activity 5.

Agenda

These activities attempt to implement the following recommendations of NCTM's *Agenda for Action* (1980):

- Coordinate mathematics with other subjects (6)
- Improve spatial perception and spatial relationships (2.3)
- Practice measuring with tools and estimating (2.3)
- Use manipulatives to illustrate concepts and develop skills (4.3)
- Take advantage of computer capabilities, particularly graphics (3)
- Encourage students to experiment (1.1)
- Accommodate diverse interests and abilities (6)
- Supply enrichment opportunities as an alternative to acceleration (6.6)

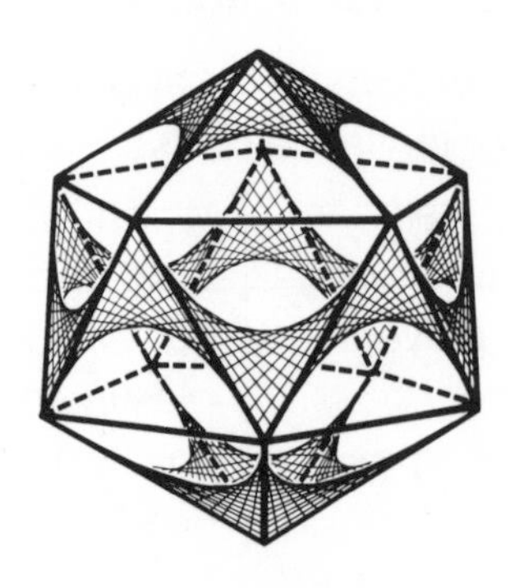

ACTIVITY 1

Raindrop in a Triangle

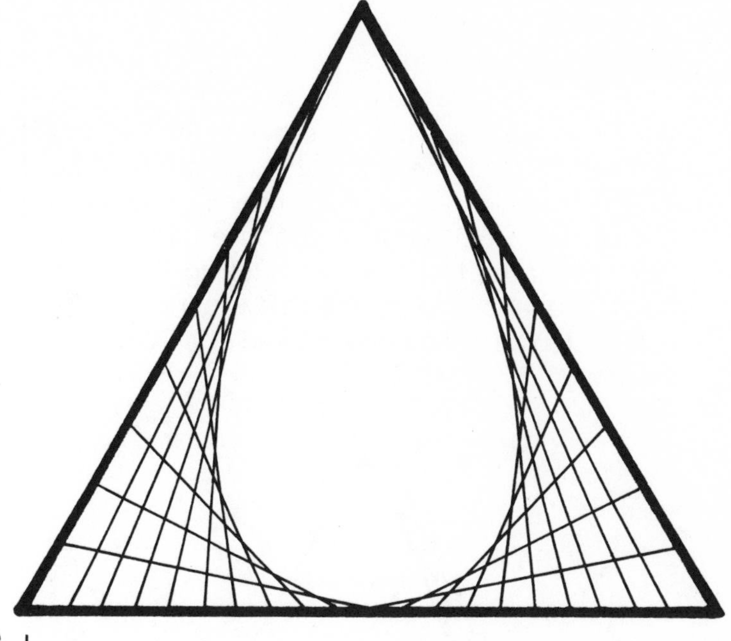

Materials

Compass
Centimeter ruler
Scissors
Pencil
Transparent tape, small amount
Cardboard, 23 cm square
String, 2.3 m

Procedure

1. *Construct an equilateral triangle.* Near the bottom of your square, and parallel to the edge, draw a line 10 cm long (fig. 1). If you do not have a centimeter ruler, use the paper ruler below. Call the endpoints of the line *P* and *Q*. Set your compass with tips 10 cm apart. Place the pointed tip of your compass on *P*, and draw an arc that crosses the line at *Q*, as shown in figure 1. Keeping the length of 10 cm on your compass, place the pointed tip on *Q* and draw another arc crossing the previous one. The point where these arcs intersect is point *R*. Connect points *Q, R,* and *P* with straight line segments. Triangle *PQR* is an equilateral triangle.

2. *Mark the slots.* With a pencil, divide two sides of the triangle into ten equal parts using the centimeter markings on a ruler. Divide the third side (the base) into twenty equal parts by marking each 1/2-cm interval. Find the middle of this third side and mark it with a cross mark (fig. 2).

3. *Cut the triangle and slots.* With scissors, cut out the triangle and then cut a slot about 0.3 cm deep at each mark. Cut these slots perpendicular to the sides. Now cut a slot at the vertex between the two sides that are marked alike. See the top slot in figure 3.

4. *Weave the first half.* Begin by tying a knot about 2 cm from the end of your string. Pull the

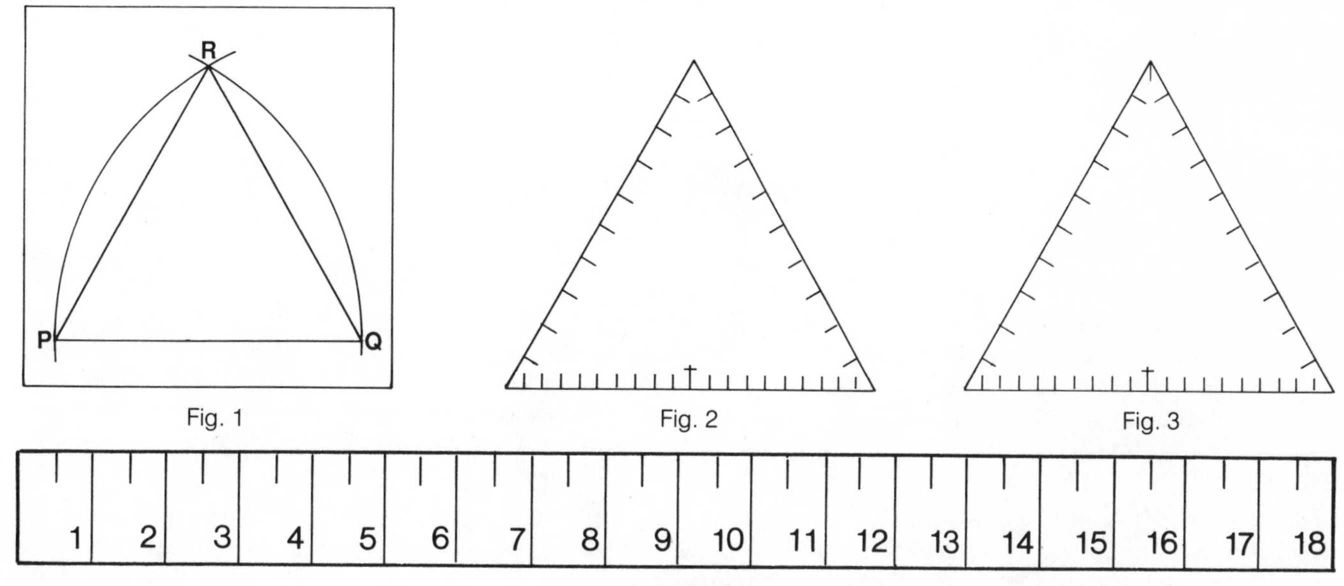

Fig. 1 Fig. 2 Fig. 3

Centimeter ruler

string through slot A (fig. 4), securing the knot underneath. Stretch the string across the top of the triangle from slot A to A′, then under the triangle to B, then above to B′, under to C, above to C′, and so forth. Finish the first half by stretching the string below to J, above to J′, and below to J again. Be sure to stop on J, the cross-marked slot. Do not cut the string.

5. *Weave the second half.* Continue the weaving by pulling the string above from J to K′ (see fig. 5), then below to K, above to L′, and so forth, ending by pulling it above from S to A′. Tie a knot below A′, and cut the string about 2 cm from the knot. To keep the string secure, tape the two ends in place. This completes the construction of a raindrop in an equilateral triangle.

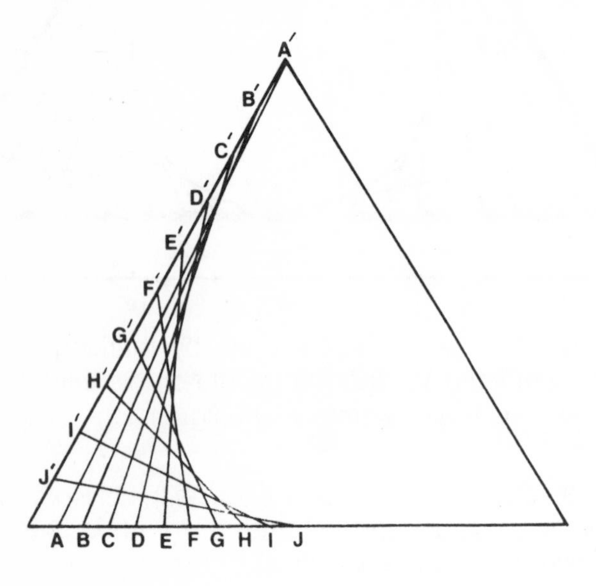

Fig. 4

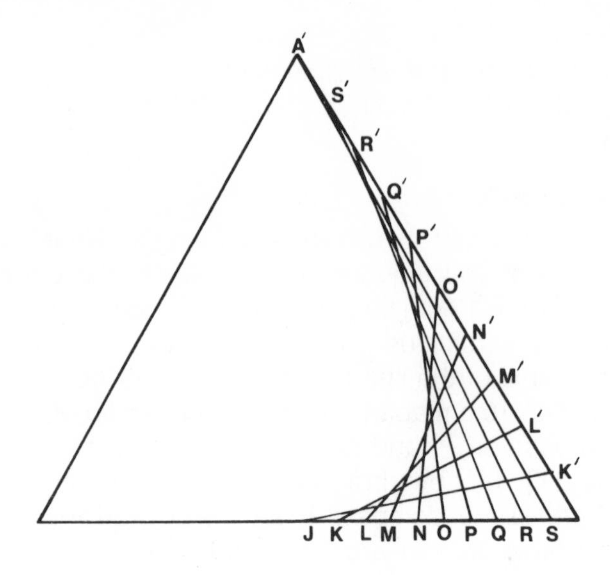

Fig. 5

TEACHER NOTES

Activity 1

Grades 5–7

Geometric concepts
Equilateral triangle
Arc and intersecting arcs
Perpendicular
Labeling triangle *PQR*
Labeling points

Skills
Using compass to draw arcs
Measuring 1 cm and 1/2 cm
Approximating 0.3 cm
Measuring 10 cm between tips of compass
Finding midpoint of segment

If students have not used a compass, use a protractor to construct the triangle as in Activity 2, or cut out triangles on a paper cutter.

Note: Each activity provides the opportunity to teach the following geometric concepts: *vertex, angle, edge* or *side, congruent angles, congruent line segments,* and *regular geometric figure.*

ACTIVITY 2

Three Curves in a Triangle

Materials

Protractor
Centimeter ruler
Scissors
Pencil
Transparent tape, small amount
Cardboard, 21 cm square
Three strings, each 2.2 m long (can be three colors)

Procedure

1. *Draw the triangle.* Near the bottom of your square and parallel to the edge, draw a line 20 cm long. If you do not have a centimeter ruler, use the paper ruler below. At both ends of this line, use your protractor to draw a 60-degree angle. Extend the sides of these angles until they meet (fig. 1). You have constructed an equilateral triangle.

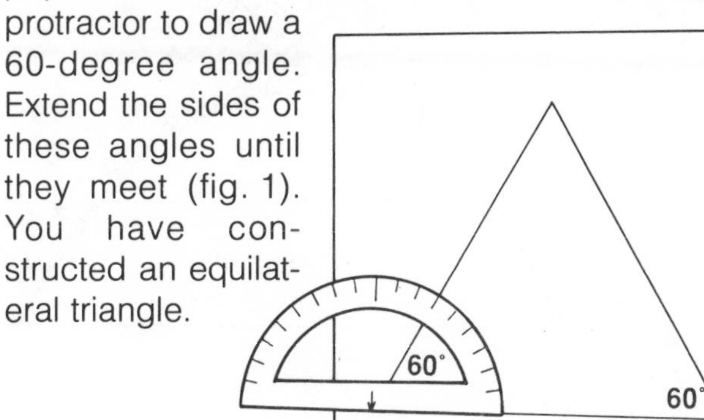

Fig. 1

2. *Mark the sides.* Use a ruler and pencil to mark 1-cm intervals on each side of the triangle (fig. 2). This should result in nineteen equally spaced marks on each side.

3. *Cut the triangle and slots.* Cut out the triangle. Next cut a slot about 1/2 cm deep at each mark. These cuts should be perpendicular to the sides of the triangle (fig. 3).

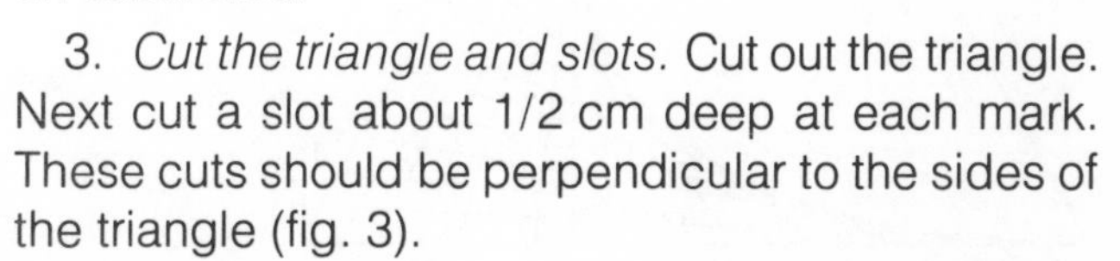

4. *Weave the first curve.* If you are using three colors of string, weave each curve with a different color. Knot the end of your first string. Following the numbering in figure 4, pull the string up through slot

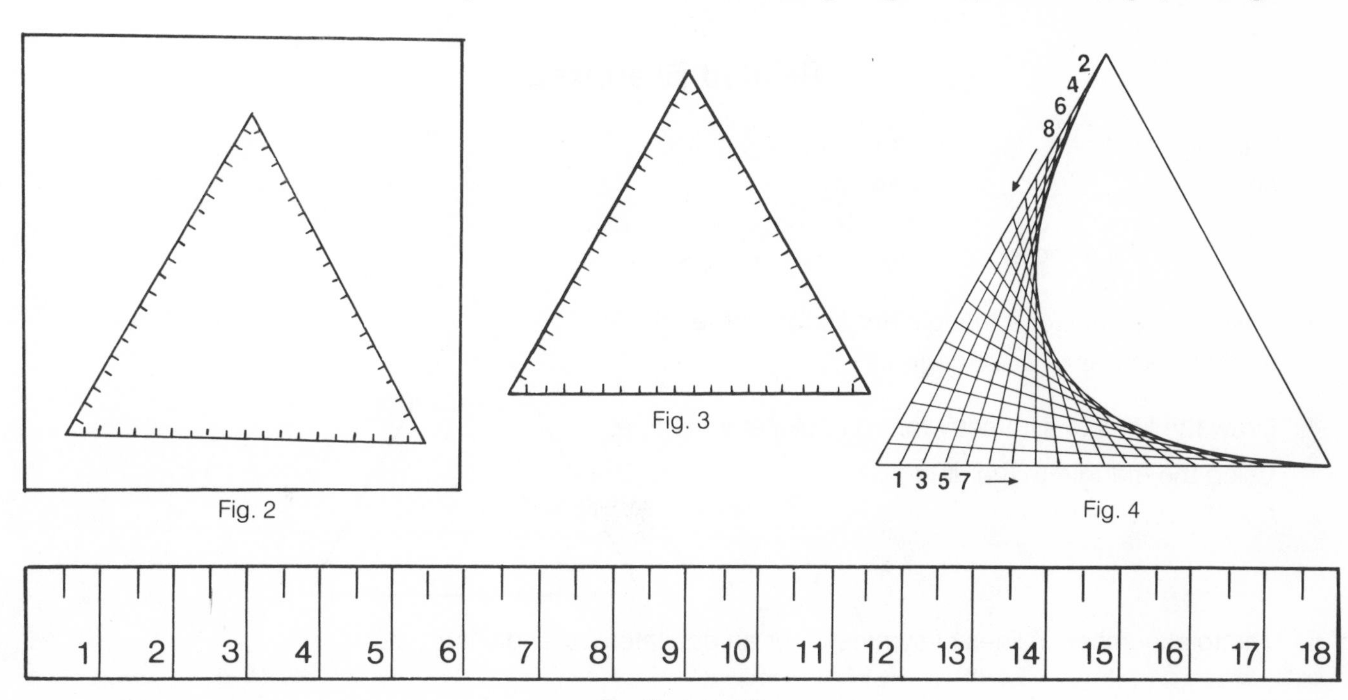

Fig. 2

Fig. 3

Fig. 4

Centimeter ruler

1, keeping the knot underneath. Then stretch the string across the top of the triangle to slot 2, underneath to slot 3, above to 4, below to 5, and so forth. Pull the string securely into each slot. When you have completed the curve, tie a knot underneath and secure the end with tape.

5. *Weave the other two curves.* Follow the numbering in figure 5 to weave the second curve. Then use figure 6 as a guide for the last curve. This completes the construction.

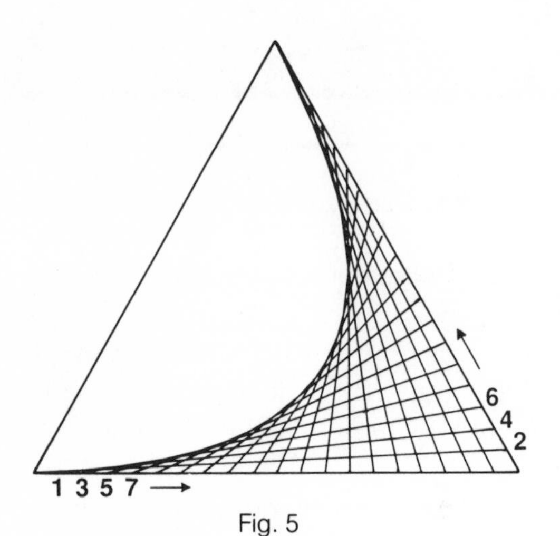

Fig. 5

Fig. 6

Related Exercises

1. A triangle that has three congruent sides is called an ____________________ triangle.

2. A triangle is a three-sided ____________________.

3. If all three angles of a triangle are congruent, each angle measures ______ degrees.

4. Draw the lines of symmetry for an equilateral triangle, using the triangle to the right.

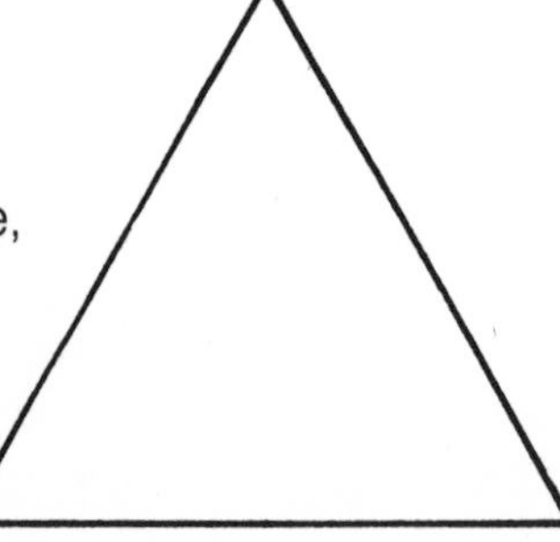

5. The total number of lines of symmetry for an equilateral triangle is ______.

Teacher Notes for Activity 2 on page 10

*ACTIVITY 3

Center of a Triangle

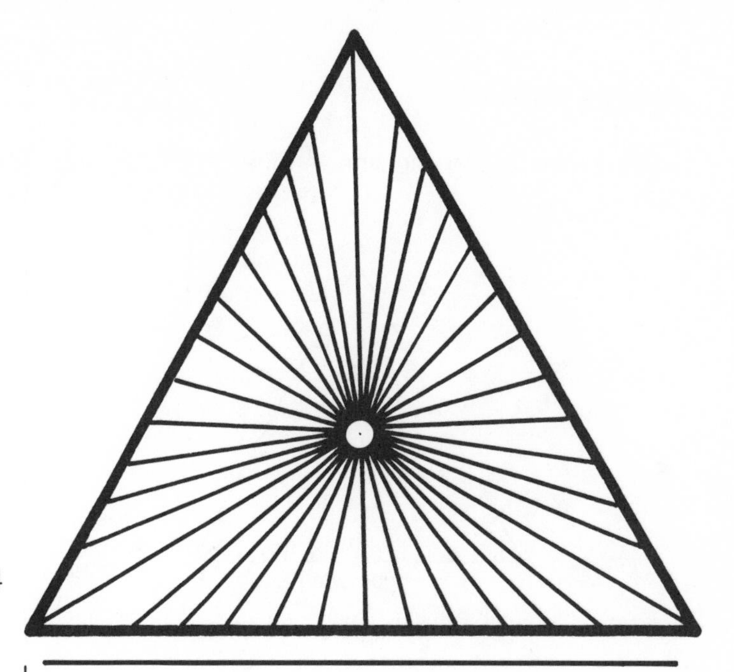

Materials

Compass
Ruler
Pencil
Scissors
Large needle
Glue
Medium-heavy crochet cotton, 16 ft.
Stiff cardboard, 5 1/2 in. square
Washer or cardboard ring with hole of diameter 1/4 or 3/8 in.
For optional steps 8 and 9: Paste
Colored paper (6 in. square) to match the string

Note: The cardboard ring can be made by punching a large hole with a paper punch and then cutting a circle around it.

Procedure

1. *Construct the triangle.* Near the bottom of your square and parallel to the edge, draw a line 5 1/4 inches long (fig. 1). Call its endpoints *P* and *Q*. Set your compass with tips 5 1/4 inches apart. Place the pointed tip of your compass on *P*, and draw an arc that crosses the line at *Q*. Keeping the length of 5 1/4 inches on your compass, place the pointed tip on *Q*, and draw another arc crossing the previous one. The point where these arcs intersect is point *R*. Connect points *Q, R,* and *P* with straight-line segments. Triangle *PQR* is an equilateral triangle.

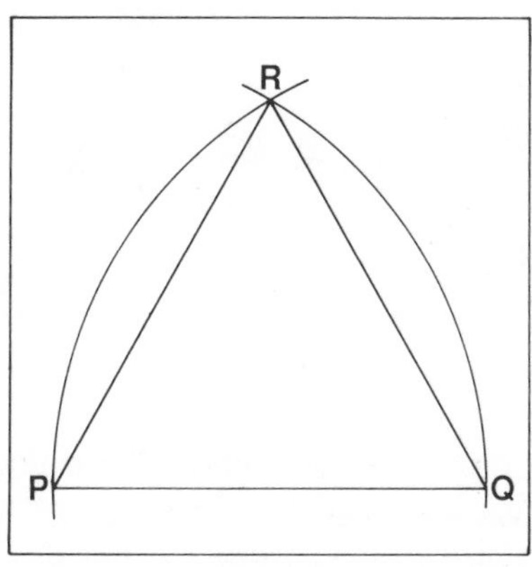

Fig. 1

2. *Mark the sides.* Use a ruler and pencil to mark 3/4 inch from each vertex on each of the three sides of the triangle. Then between the two marks on each side, mark every 3/8 inch (fig. 2).

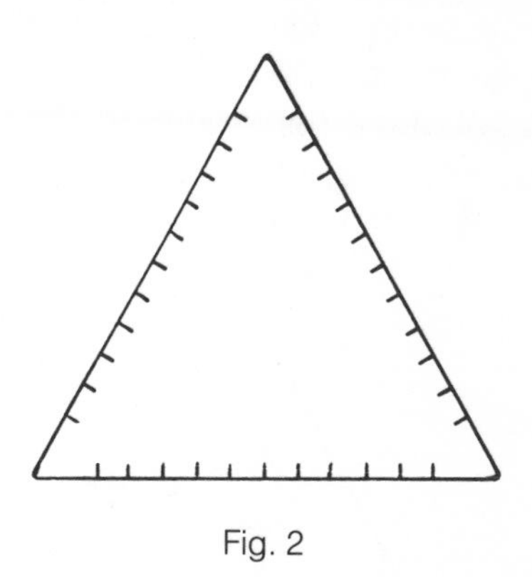

Fig. 2

3. *Locate the center.* Find the midpoint of two different sides. The midpoint is the sixth mark from each end. Draw a line from each midpoint to its opposite vertex (fig. 3). The point where these two lines cross is the center of the triangle. Connect the center to the third vertex.

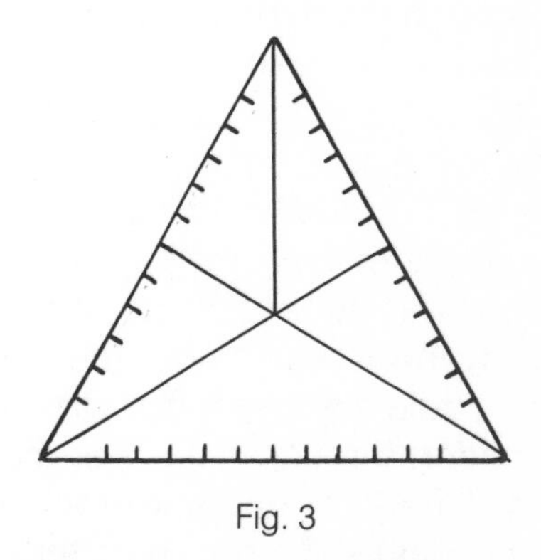

Fig. 3

4. *Cut the triangle.* Cut out the triangle. Then at each mark cut a slot deep enough to hold two strings (1/8 to 1/4 inch). Also cut a 1/4-inch slot at

each vertex along the line leading to the center (fig. 4).

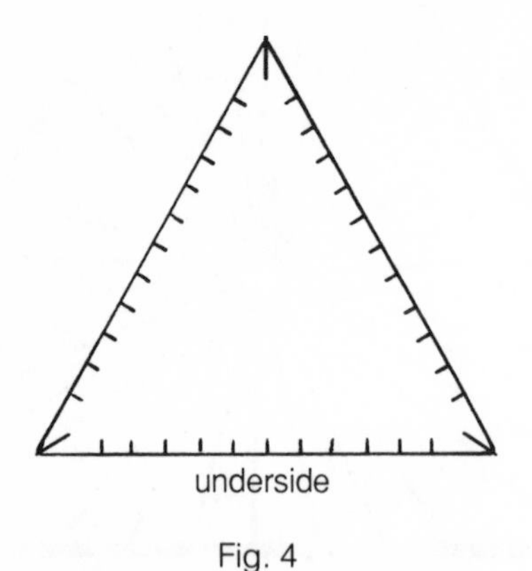

Fig. 4

5. *Mark the center.* On the unmarked side of the triangle, which will be the top of the design, mark a light pencil dot at the center (fig. 5). This point can be found by applying pressure on the center underneath with a pin or any pointed object.

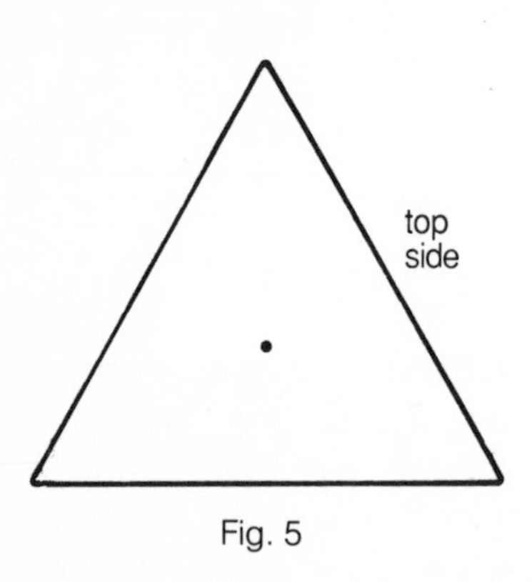

Fig. 5

6. *Position the ring.* Thread the needle with 4 feet of string and tie a knot at the end. Next pull the thread up through the middle slot on any side, securing the knot underneath. Place the ring on the triangle so the dot is in the center of the ring. This dot must stay in the center of the ring throughout the weaving. Bring the needle through the ring from underneath and then over the top of the ring and back through the same slot, pulling the thread tightly. Stretch the thread under the triangle and through the middle slot of another side. Again pull the thread under, through, and over the ring and back to the slot. Perform the same procedure for the third side. In figures 6 and hereafter, each line through the ring represents two strings. If necessary, go back over your work until all three strings are tight and the dot is in the center of the ring.

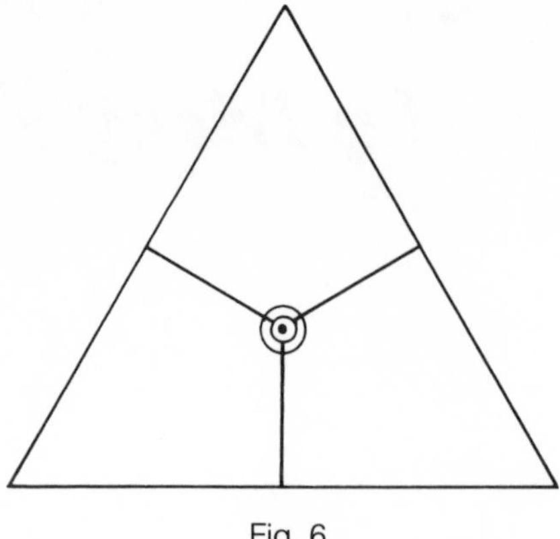

Fig. 6

7. *Weave the design.* Now pull the thread through all the slots, one by one. Each time, draw the thread under, through, and over the ring and back to the slot. Then stretch the string under the triangle to another slot. It helps to tilt the ring to get the needle under it. Strings should be side by side on the ring with none overlapping. The strings will cover the ring. Carefully push them together to create needed space. If the tension of the strings begins to pull the ring off center, use some of the slots on the opposite side of the triangle (fig. 7). When you are out of string, tie a knot under the triangle. Then begin another 4 feet of string with a knot at the end. Proceed until you have passed the string through every slot. Tie a knot underneath and cut off the extra string.

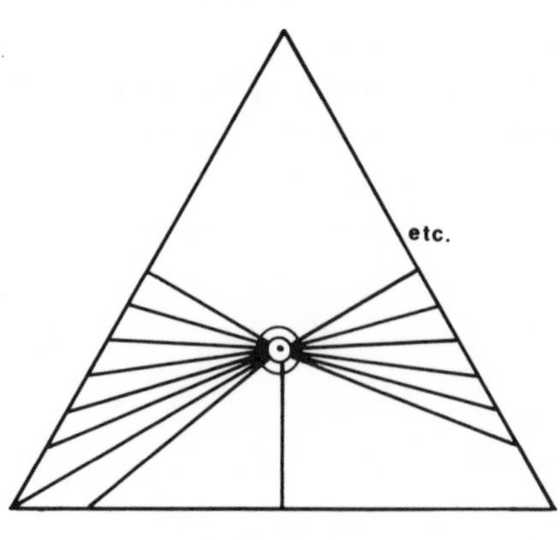

Fig. 7

8. *(Optional) Draw a frame.* Place your triangle upside down on the colored paper and trace around it. Then put the triangle aside and with a pencil and ruler extend each side (fig. 8a). Next, on each extended side measure and mark 1/4 inch (fig. 8b). Join the points to form flaps (fig. 8c). Through each vertex of the triangle, draw a line that cuts off the tips of the flaps (fig. 8d).

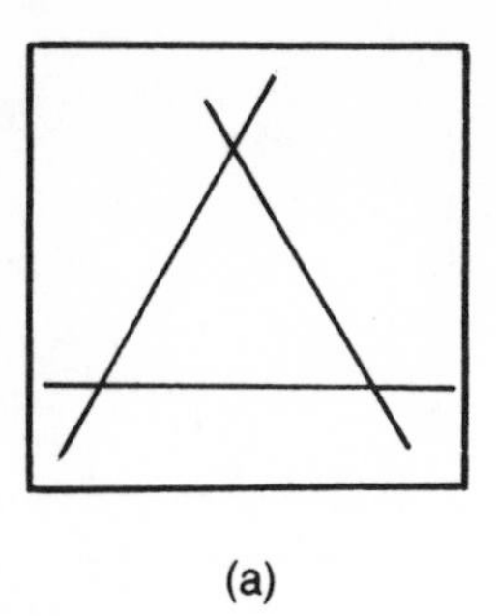

(a)

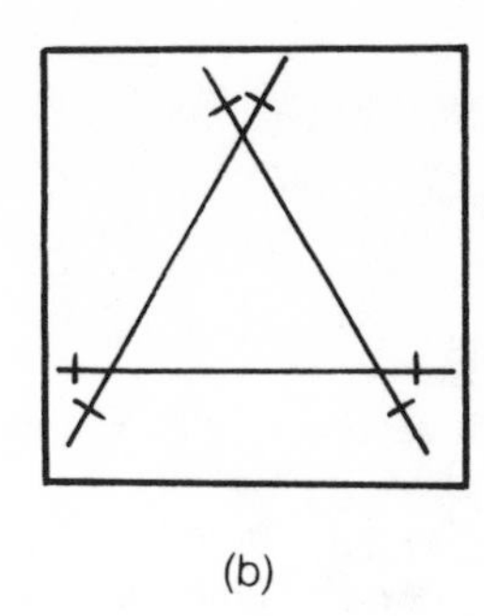

(b)

(c)

(d)

Fig. 8

9. *(Optional) Cut, fold, and paste the frame.* Cut out the triangle with its unpointed flaps attached (fig. 9). Glue the original triangle on the inner triangle, making the sides fit (fig. 10). Fold the flaps forward and glue them down onto the edges of the original triangle to form the frame.

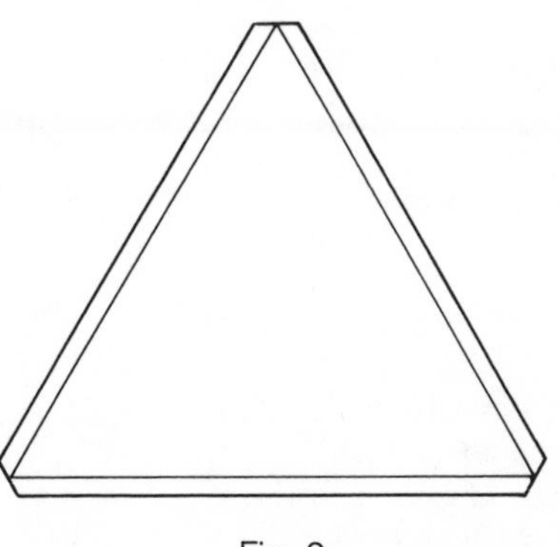

Fig. 9

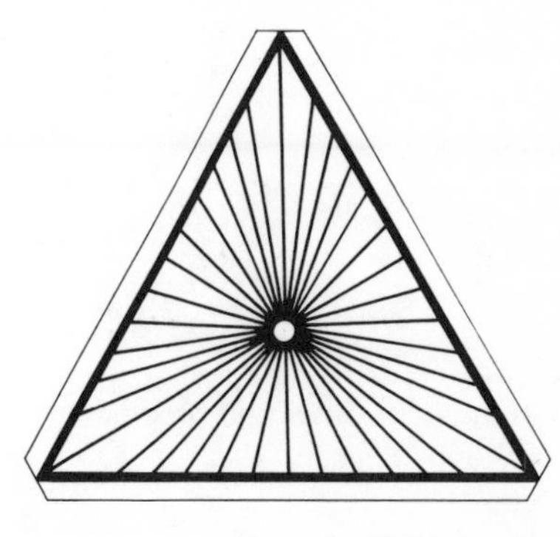

Fig. 10

Related Exercises

1. Draw an inscribed circle in one of the triangles to the right and then a circumscribed circle about the other triangle.

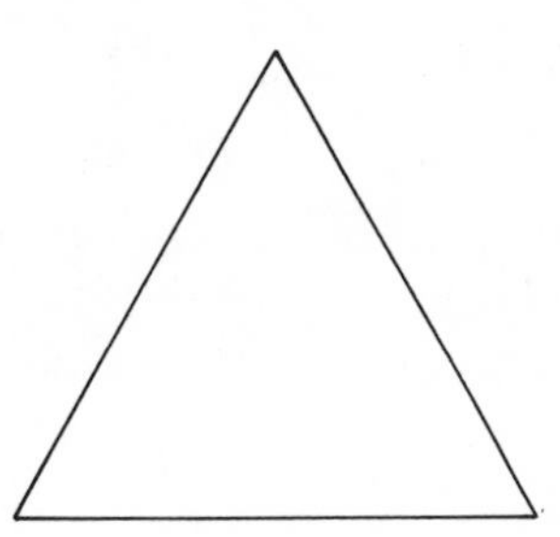

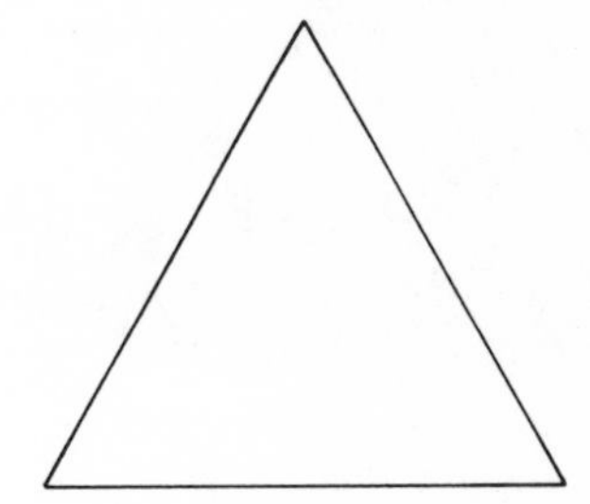

*2. The point where the bisectors of the angles of a triangle meet is called the ____________________ ____________ of the triangle.

*3. The point where the perpendicular bisectors of the sides of a triangle meet is called the __________ ____________ of the triangle.

*4. The center of a triangle (see step 3 of Activity 3) is actually the center of both the inscribed circle and the circumscribed circle of the triangle. This center was found by bisecting two angles of the triangle. Can you prove that the two intersecting lines really are angle bisectors?

Teacher Notes for Activity 3 on page 10

TEACHER NOTES

Activity 2

Grades 5–7

Geometric concepts
60° angle
Equilateral triangle
Perpendicular
Line of symmetry

Skills
Measuring 2.2 m of string
Using a protractor to draw 60° angles
Using a ruler to mark off 1/2 cm intervals
Drawing a 20-cm segment
Approximating 1/2 cm

Answers to Related Exercises

1. equilateral
2. polygon
3. 60°
4.
5. 3

TEACHER NOTES

*Activity 3

Grades 8–10

Geometric concepts
Center of triangle (incenter and circumcenter)
Midpoint
Opposite side
Arc and intersecting arcs

Skills
Measuring 3/4 and 1/4 in. from a point
Drawing marks 3/8 in. apart
Using compass to draw arcs
Approximating 1/8 and 1/4 in.
Measuring 5 1/4 in. between tips of compass
Threading a large needle

Answers to Related Exercises

1.  
2. incenter
3. circumcenter

ACTIVITY A

On Your Own

Triangular Gem in a Triangle

Based on Activities 1 and 4

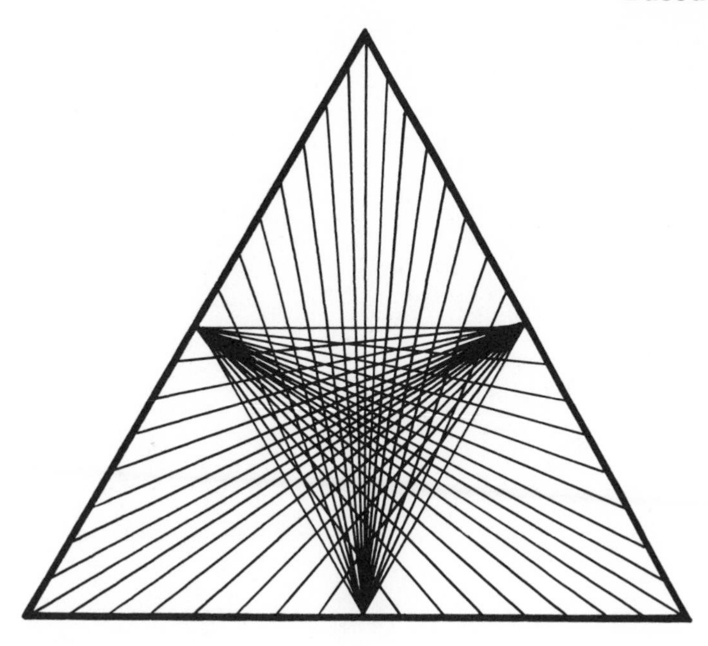

This little gem has a construction very similar to that of the star. Ask your teacher for a copy of Activity 4 or 7. If you study one of them, you should be able to mark this triangle and weave the design without step-by-step instructions. Choose whatever size of triangle you prefer.

TEACHER NOTES

Activity A

Grades 5–7

See Activity 1 or 2 for triangle and Activity 4 for string design.

ACTIVITY 4

Star in a Square

Materials

Centimeter ruler
Pencil
Scissors
String, 9.7 m
Cardboard, 12 cm square

Procedure

1. *Draw a square.* First of all be sure that your 12-cm square has four 90-degree angles and that each side is 12 cm long. Figure 1 shows the steps for drawing a 12-cm square.

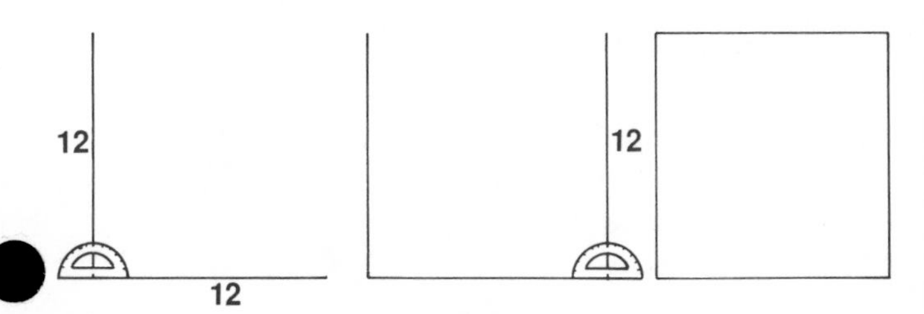

Fig. 1

2. *Prepare the square for weaving.* With a pencil, divide each side of the square into twelve equal parts by using the centimeter markings on a ruler. If you do not have a centimeter ruler, use the paper ruler below. Next find the middle of each side and mark it with a V as shown in figure 2. With scissors, cut a slot about 1/2 cm deep at each mark, and cut a V-shaped notch (at least 1/2 cm deep) at each of the four midpoints.

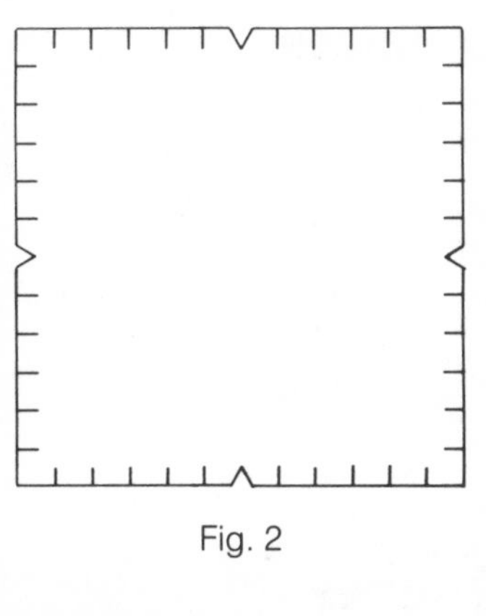

Fig. 2

3. *Weave one point of the star.* Begin by tying a large knot near the end of the string. Pull the string up through slot 1 (fig. 3) with the knot under the square. Stretch the string across the top of the square from slot 1 to the midpoint of the opposite side, then underneath to slot 2, above to the midpoint, below to 3, and so forth. Do not cut the string when you have finished the first point of your star.

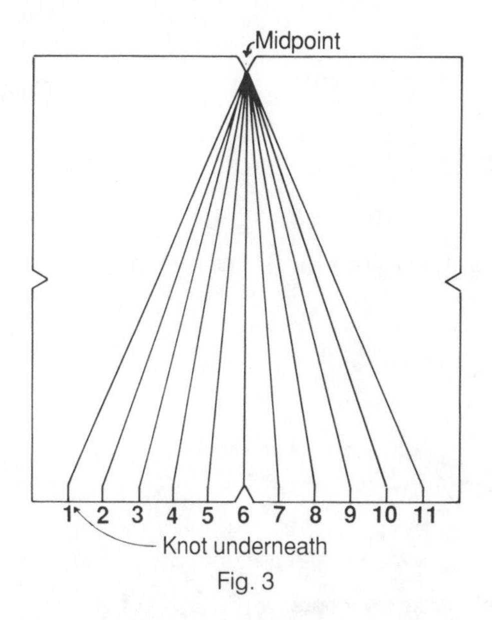

Fig. 3

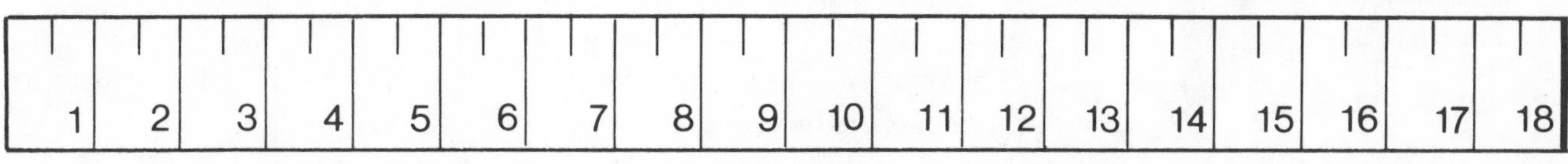

Centimeter ruler

4. *Weave the second point of the star.* Pull the string underneath from the finished midpoint to slot 1 of an adjacent side (fig. 4). Now proceed as in step 3. That is, stretch the string across the top from slot 1 to the midpoint of the opposite side, below to slot 2, above to the midpoint, and so forth.

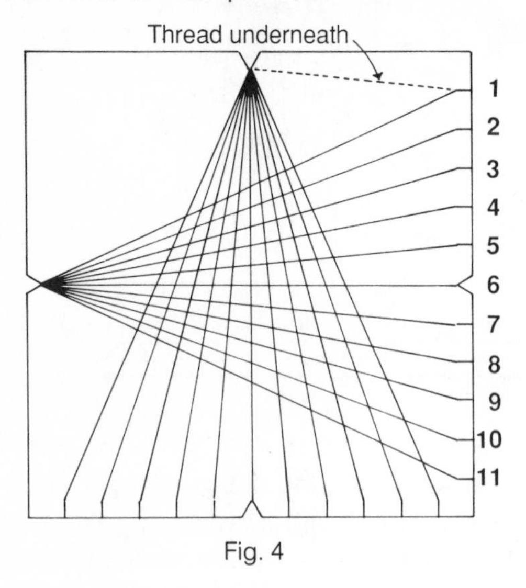

Fig. 4

5. *Weave the last two points of the star.* For both of the remaining points, follow the same procedure as in step 4 but skip slot 6, since you already have a string in that position. See figure 5. After you have completed the star, tie the string to any other string on the underside. Cut off the extra string. This completes the construction.

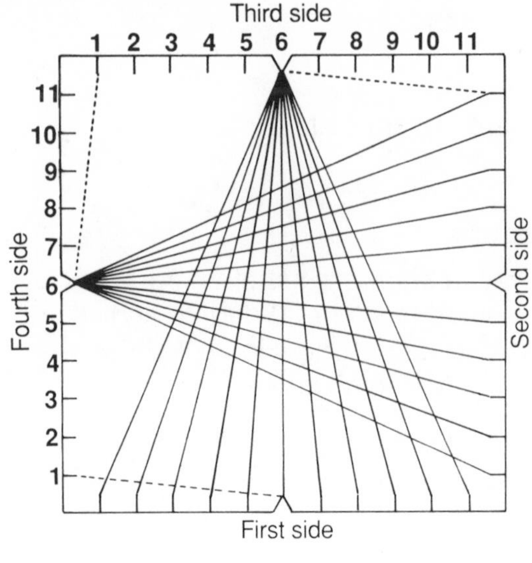

Fig. 5

Related Exercises

1. 10 ____________________ = 1 cm
2. ______ cm = 1 m
3. Another name for 9.7 m is 9 m ______ cm

TEACHER NOTES

Activity 4

Grades 4–6

Geometric concepts
Square (90° angles and congruent sides)
Opposite side
Adjacent side

Skills
Measuring centimeters
Measuring 9.7 m of string
Approximating 1/2 cm
Finding midpoint

Answers to Related Exercises

1. mm, or millimeters
2. 100
3. 70

ACTIVITY 5

Two-Layer Design on a Square

Materials

Centimeter ruler
Pencil
Scissors
Cardboard, 10 cm square
Transparent tape, small amount
Thin string, 2.7 m
Heavier string, preferably of a second color, 3.6 m

Procedure

1. *Mark and cut the slots.* With a pencil, mark each 1/2 cm on all sides of the square. If you do not have a centimeter ruler, use the paper ruler below. Then with scissors cut a slot about 1/2 cm deep at each mark (fig. 1).

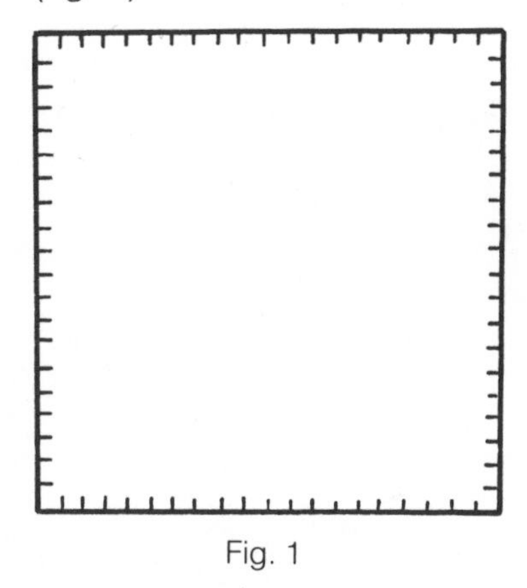

Fig. 1

2. *Weave the first layer.* For the weaving in steps 2 and 3, keep the string on a spool and unwind it as you need it. Use the thin string first. Tie a knot about 3 cm from the end of the string and then follow the numbering in figure 2. Bring the string up through slot 1, and secure the knot underneath with a bit of tape. Next, stretch the string across the top to slot 2. On the underside, bypass one unnumbered slot and pull the string up through slot 3. Notice that you skip a slot each time. Stretch the string across the top to 4, below to 5, and so forth. Continue this process using every other slot until you pull the string down through the end slot marked A. Then bring the string underneath the corner and up through slot B, across the top to C, bypass a slot underneath to D, and so forth. When you have completed the design pictured in figure 2, cut off most of the extra string and tape a short end underneath.

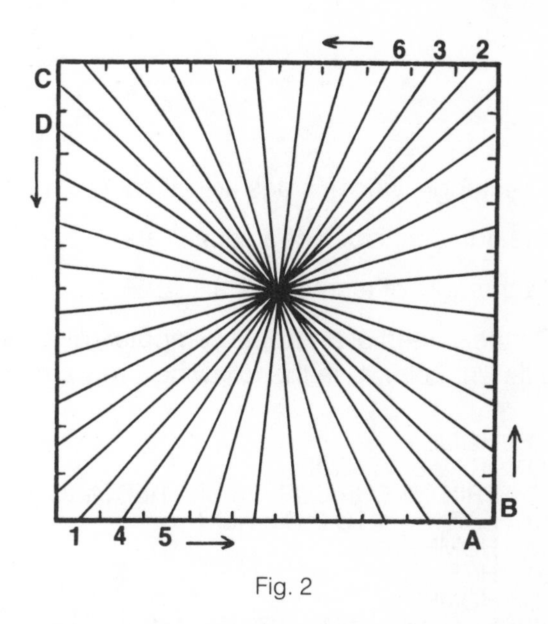

Fig. 2

3. *Weave the second layer.* Use the heavier string for this upper design. Tie a knot about 3 cm

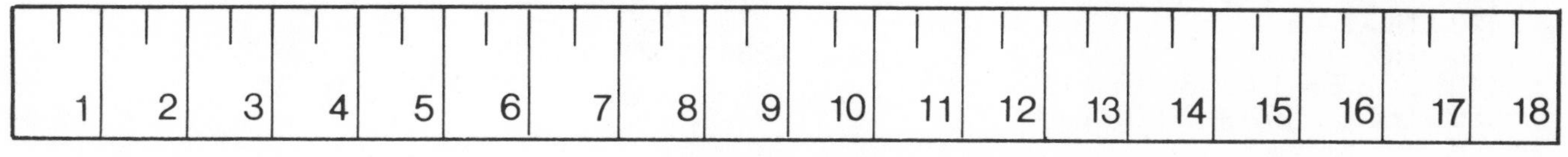

Centimeter ruler

from the end of the string. Bring the string up through slot 1 (see fig. 3) and secure the knot underneath with a bit of tape. Then, following figure 3 and using *every slot* this time, stretch the string across the top of the square to slot 2, underneath to slot 3, above to 4, below to 5, and so on. When you have completed the curve on the first two sides, pull the string under the corner as in step 2. Then weave an identical curve on the other two sides of the square, beginning with B to C to D. After you have pulled the string through the last slot, cut off any extra string and tape a short end underneath.

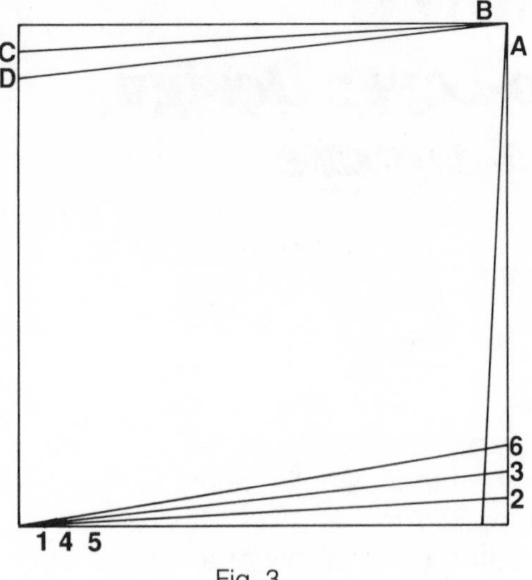

Fig. 3

Related Exercises

1. Draw the lines of symmetry for a square.

2. A rectangle whose sides are all congruent is called a ____________________.
3. The adjacent sides of a square are ____________________.
4. Each angle of a square is a ____________________ angle.
5. On the screen of an Apple computer, display four curves, one in each corner of a rectangle, each in a different color. Use the following program, written by Karl West and adapted by the author.

```
10  HOME
20  HTAB 8: VTAB 13
30  PRINT "FOUR CURVES IN A RECTANGLE"
40  FOR X = 1 TO 2500: NEXT
50  HOME
60  HGR
70  HOME
80  HCOLOR= 1:D = 279:E = 159
90  FOR A = 1 TO 279 STEP 10
100  LET C = INT (E / D * A)
110  HPLOT A,B TO D,C
120  NEXT A
130  HCOLOR= 5
140  FOR B = 1 TO 159 STEP 10
150  LET C = 279 - INT (D / E * B)
160  HPLOT D,B TO C,E
170  NEXT B
180  HCOLOR= 2
190  FOR F = 279 TO 1 STEP - 10
200  LET C = INT (E / D * F)
210  HPLOT F,E TO O,C
220  NEXT F
230  HCOLOR= 6
240  FOR G = 159 TO 1 STEP - 10
250  LET C = 279 - INT (D / E * G)
260  HPLOT O,G TO C,O
270  NEXT G
280  END
```

Teacher Notes for Activity 5 on page 26

***ACTIVITY 6**

Circle in a Square

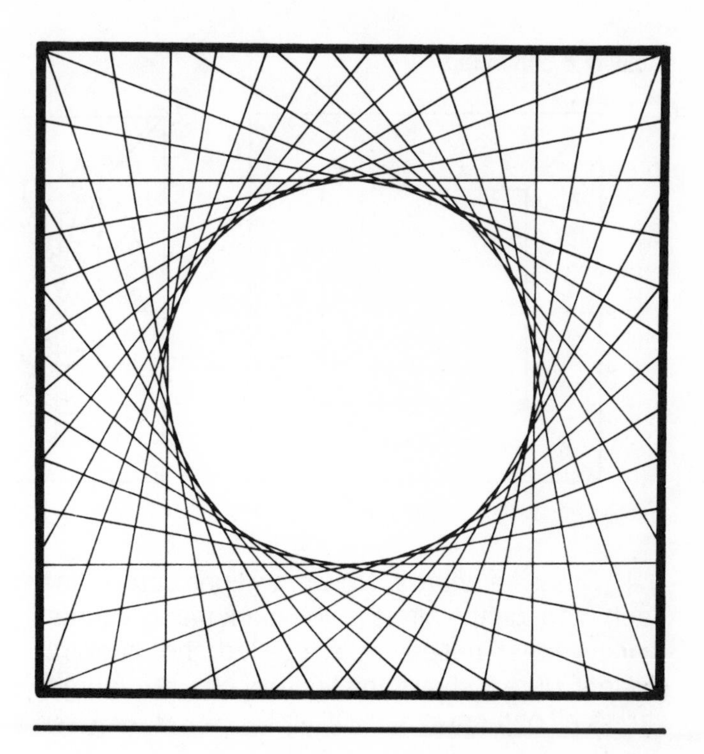

Materials

String, 22 ft.
Compass
Protractor
Ruler
Pencil
Scissors
Cardboard square, 6 1/2 in. on a side
Colored paper, 4 strips, each 1/4 in. wide and 4 in. long (Use to frame the design.)

Procedure

All the construction marks for this activity are to be done on the top side of the cardboard. Use pencil, since some marks must be erased later.

1. *Draw a square.* Use cardboard and follow figure 1 to make a square with sides 6 1/2 inches long.

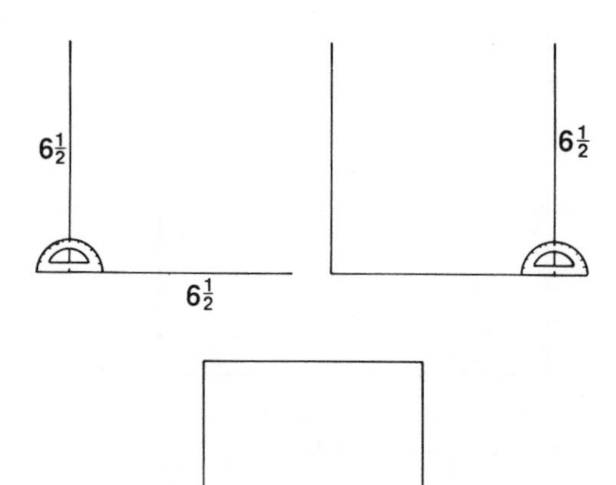

Fig. 1

2. *Construct a smaller square.* Figure 2 shows the procedure to be followed here. With light penciling and your ruler, draw lines connecting opposite vertices of the first square. Now use the point where the two lines intersect as a center and with your compass draw a circle with a radius of 2 3/4 inches. With very dark lines or red lines draw a square by connecting the points where the circle and lines intersect.

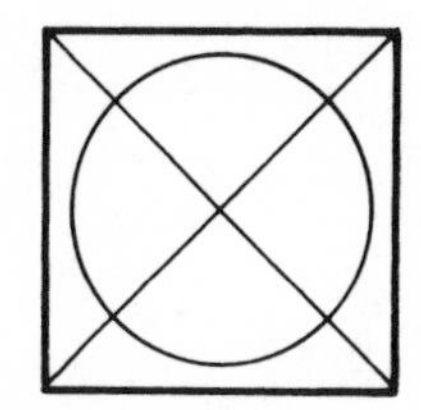
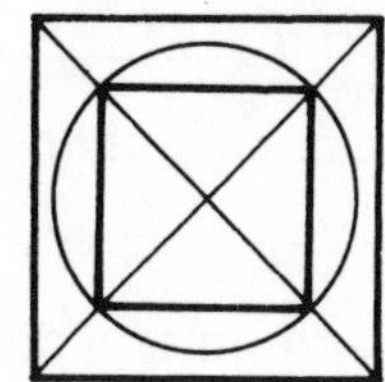

Fig. 2

3. *Divide the circle into thirty-six congruent arcs.* Use your protractor and light penciling to mark off on the circle central angles of measure 10°. Begin at 0° and end at 170°, as shown in figure 3. Now use these marks and the center of the circle to divide the circle into thirty-six congruent arcs. Figure 4 shows how to use each mark, the center, and a ruler to mark the other side of the circle. After you have penciled the thirty-six marks on the circle, erase all marks inside the smaller square.

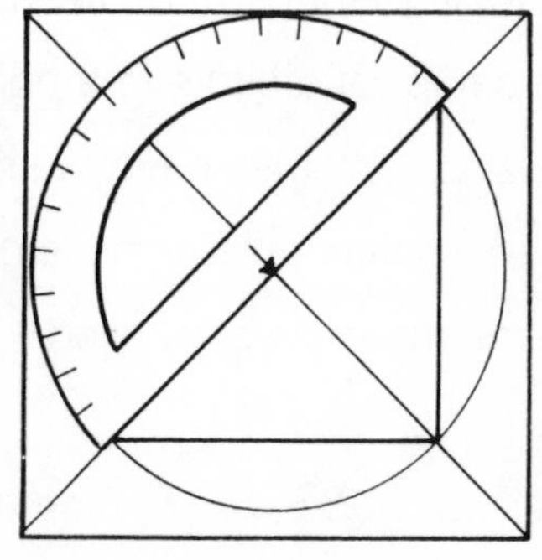

Fig. 3

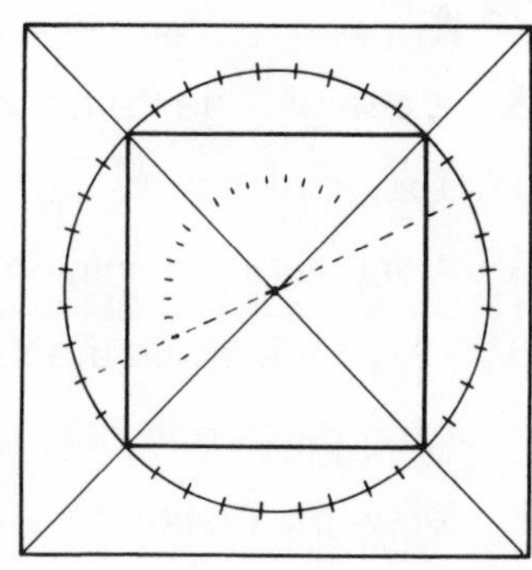

Fig. 4

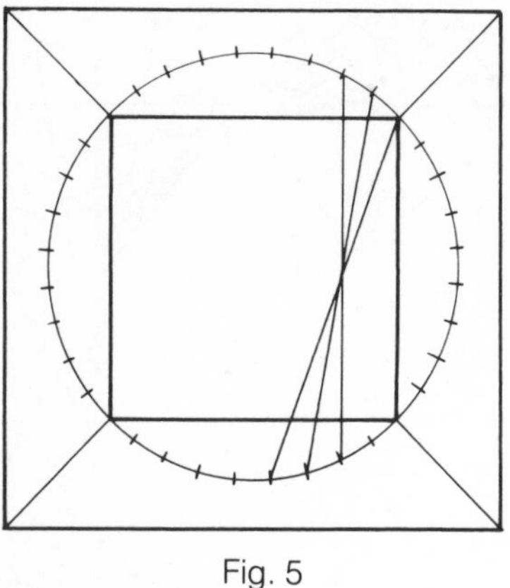

Fig. 5

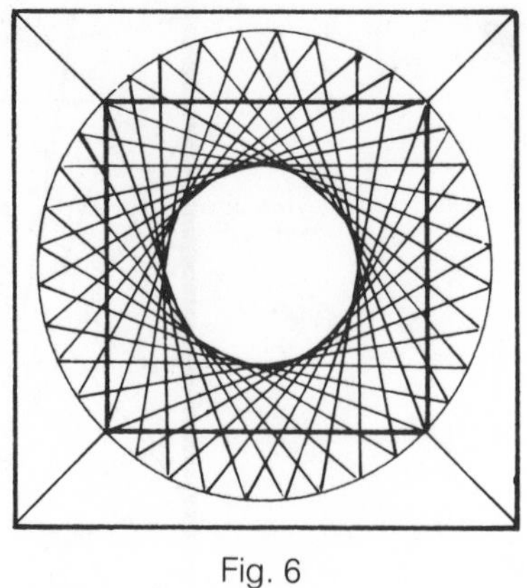

Fig. 6

Fig. 7

4. *Draw the chords.* Now draw lines connecting each mark on the circle to the mark that is thirteen marks away. These lines are called chords of the circle. Figure 5 shows the first few chords; figure 6 shows all of them.

5. *Cut the square and slots.* Cut out the small square and discard the outer square. Next cut slots about 1/8 inch deep at the ends of each line on the outer edge of the small square (see fig. 7). Since the lines are slanted, some slots may intersect each other and even cause a tiny wedge to be cut away. These little flaws will be covered by the frame later.

6. *Weave the circle.* Tie a knot about one inch from the end of your string. Draw the string through any slot, securing the knot on the underside with a bit of tape. (Remember, the top side is the one with lines.) Stretch the string across the top of the square so that it covers a line. In general, you need to cover all the lines with string, and you may do it in any order. The numbering in figure 8 is one possibility. It causes the strings to overlap in an attractive, orderly fashion. To follow figure 8, begin at slot 1, stretch the thread across the top to slot 2, pull it underneath to 3, above to 4, below to 5, and so on. Continue the process of choosing the next line that slants slightly more than the previous line until you have used all the slots. The circle will then be complete.

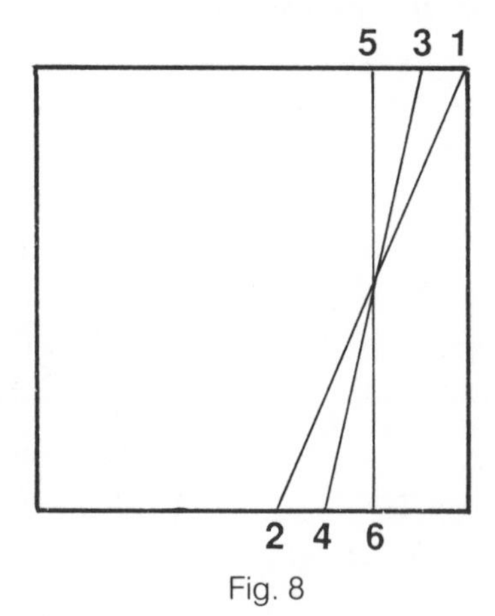

Fig. 8

Related Exercises

1. A ____________ is an instrument used to draw arcs and circles.
2. A ____________ is an instrument used to measure angles.
3. A straight line from the center of a circle to a point on the circle is called a ____________ .
4. Two curves are ____________ if they have the same size and shape.
5. A line segment joining two points on a circle is called a ____________ .
6. Two angles with the same measure are called ____________ angles.
7. An arc is a part of a ____________ .
8. Draw the design for Activity 1, 2, 5, 8, or 10 on the screen of a computer.

Teacher Notes for Activity 6 on page 26

ACTIVITY 7

Star in a Pentagon

(Using concurrent lines)

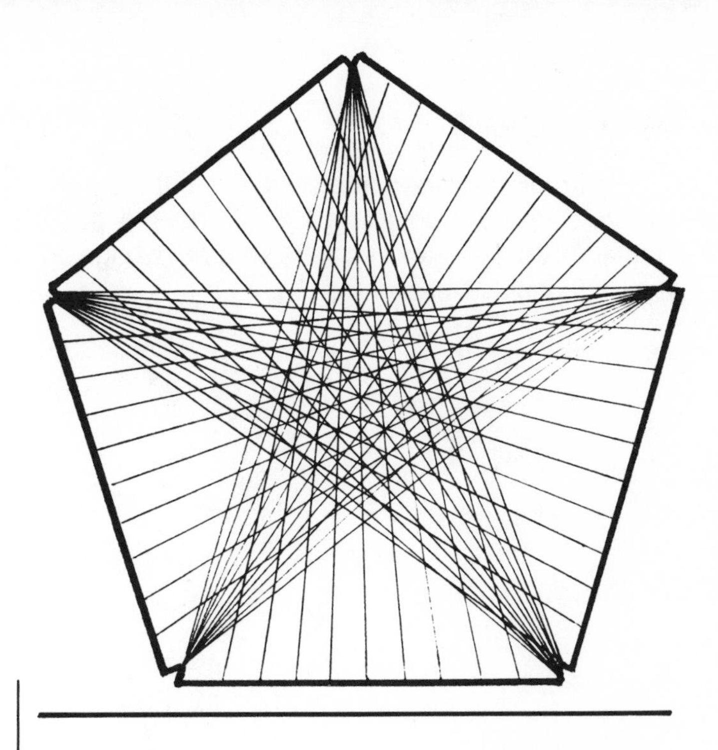

Materials

String, 11 yd.
Compass
Protractor
Scissors
Pencil
Transparent tape, small amount
Cardboard square, 6 in. on a side

Procedure

1. *Construct the pentagon.* Find the point that is approximately the center of the square. Using this point as one end, draw a 2 1/8-inch line perpendicular to a side of the square (fig. 1). With your compass, draw a circle using this line as a radius.

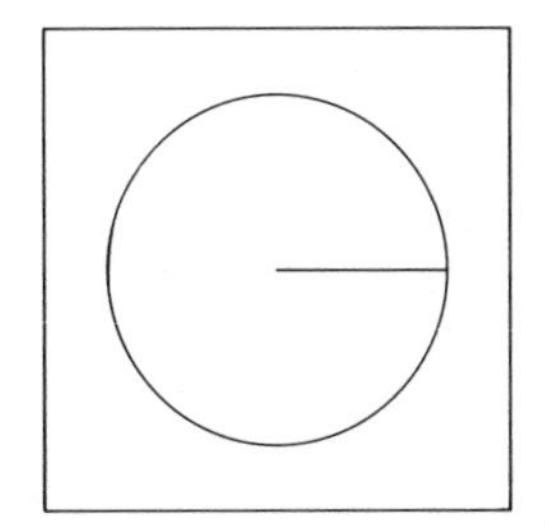

Fig. 1

Next, with your protractor, draw five adjacent 72-degree angles with vertices at the center. Figures 2 and 3 show how this can be done by first placing the protractor along the radius on one side and then placing it along the radius on the other side. With a ruler, extend the sides of the angles to the circle. Join the ends of the radii to form a regular pentagon (fig. 4).

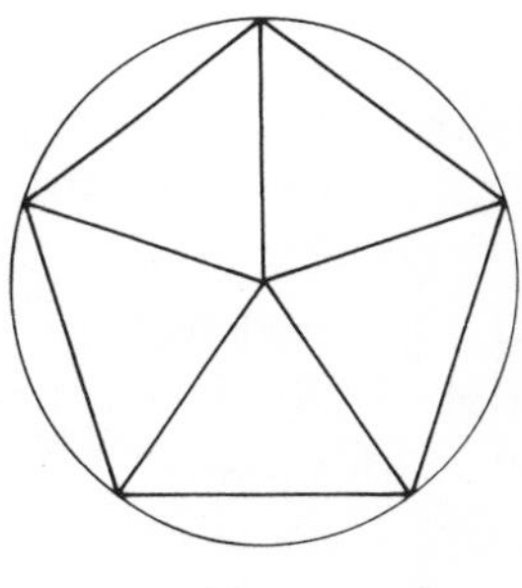

Fig. 4

2. *Mark the slots.* With a pencil, divide each side into ten equal parts by using the 1/4-inch markings on the ruler. If the sides do not come out quite even, leave them a little uneven at the ends. Next mark each vertex with a V as shown in figure 5.

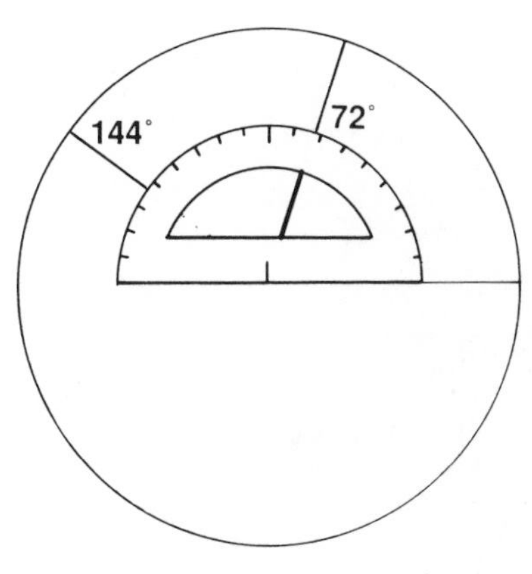

Fig. 2

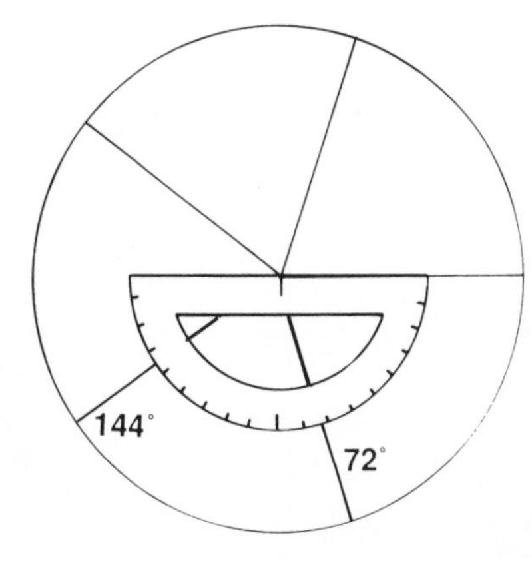

Fig. 3

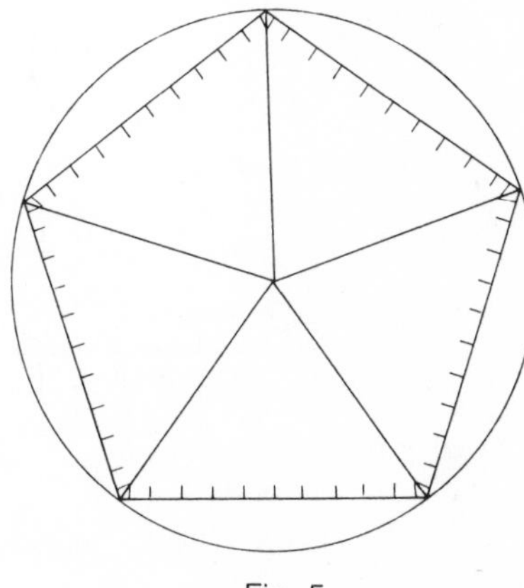

Fig. 5

3. *Cut the pentagon and slots.* Cut out the pentagon. Then cut a slot about 1/8 inch deep at each mark. At each vertex, cut a V-shaped wedge at least 1/8 inch deep (fig. 6).

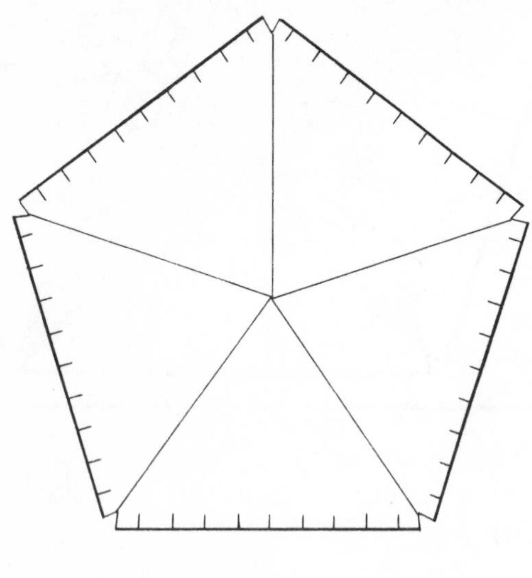

Fig. 6

4. *Weave one point of the star.* Tie a large knot near the end of the string. Pull the string up through slot 1 (on side *C D*) with the knot underneath. Stretch the string across the top of the pentagon from slot 1 to vertex *A* (fig. 7), then underneath to slot 2, above to *A*, under to 3, and so on, until you pull the string above from *D* to *A*. Do not cut the string.

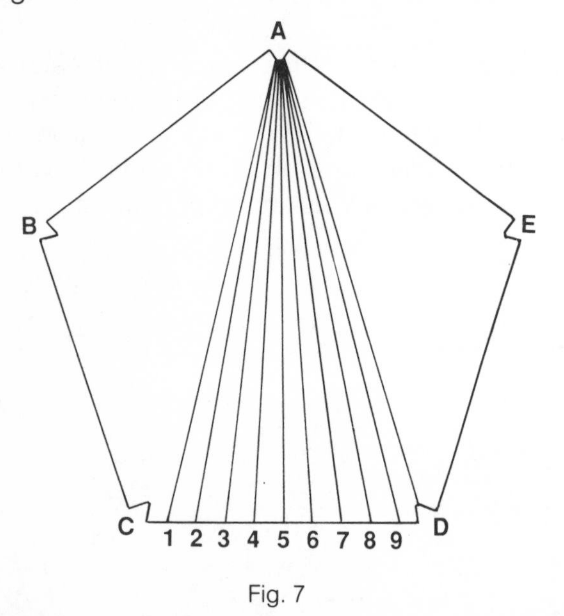

Fig. 7

5. *Weave the second point of the star.* Now pull the thread underneath from vertex *A* to slot 1 of side *D-E*, then above to vertex *B*, under to slot 2, above to *B*, and so forth, ending the point with the thread passing above from *E* to *B* (fig. 8).

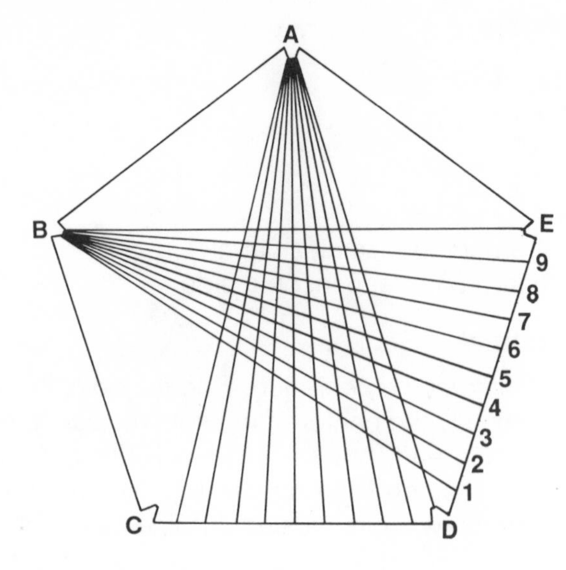

Fig. 8

6. *Weave the other three points of the star.* Pull the thread underneath from vertex *B* to the first slot of side *E-A*. Proceed as in step 5, to connect side *E-A* to vertex *C*, then side *A-B* to vertex *D*, and side *B C* to vertex *E* (fig. 9). After completing the fifth point of the star, cut the string about an inch from the cardboard and tape both ends of the string underneath. The star you have constructed is formed of concurrent, or converging, lines.

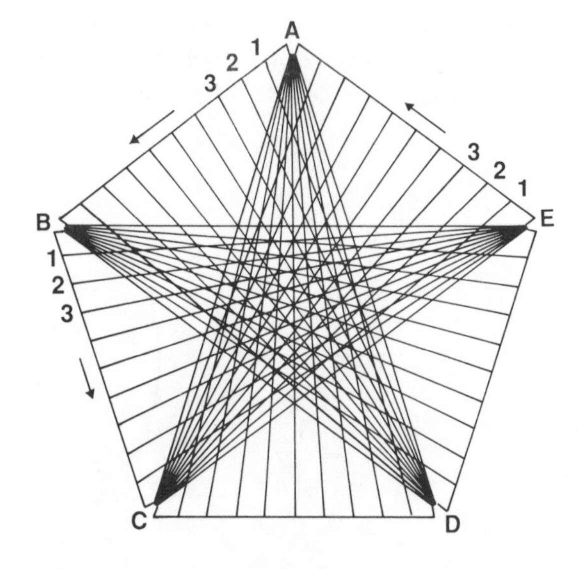

Fig. 9

Related Exercises

1. A polygon having five sides is called a ____________ .
2. If a circle is divided into five equal parts, each central angle has a measure of ______ degrees.
3. The total measure of the angles all the way around a point is ______ degrees.
4. The central angles used in constructing a pentagon are called ____________ angles because they are side by side.
5. An angle having its vertex at the center of a circle is called a ____________ angle.
6. A point where two sides of a polygon meet is called a ____________ .

TEACHER NOTES

Activity 7

Grades 6–9

Geometric concepts
Pentagon
Central angle
72° angle
Adjacent angles
Measure of a complete revolution
Equal central angles with equal arcs and chords

Skills
Using compass to draw circle
Drawing 72° adjacent angles with protractor
Approximating 1/8-in. depth for slots
Using 2 1/8 in. for radius
Drawing marks 1/4 in. apart

Answers to Related Exercises

1. pentagon
2. 72
3. 360
4. adjacent
5. central
6. vertex

*ACTIVITY B

On Your Own

Center of a Pentagon

Based on Activities 3 and 7

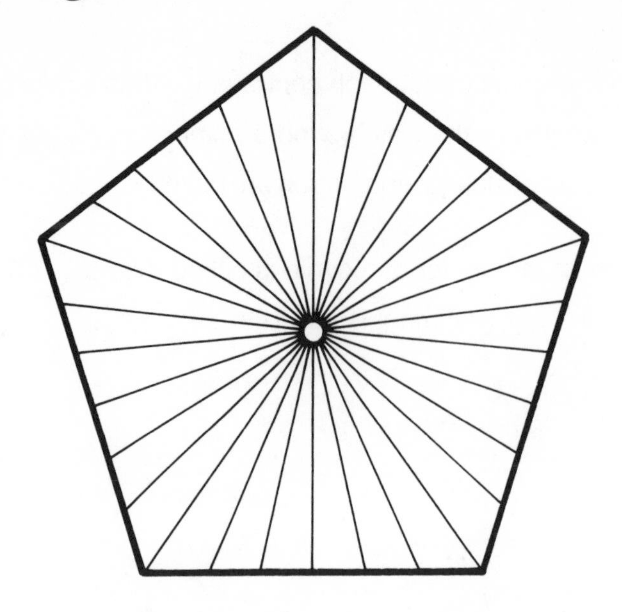

Every regular polygon has a center. Accentuate the center of a regular pentagon with string on a ring.

TEACHER NOTES

***Activity B**

Grades 8–10

See Activity 7 for pentagon and Activity 3 for string design.

***ACTIVITY 8**

Star in a Pentagon (Using parallel lines)

Materials

Compass
Protractor
Ruler
Pencil
Paste or glue
Tape, small amount
Heavy cardboard, 7 in. square
String (preferably variegated), 10 to 12 yd.
Scissors

Note: Use contrasting colors for string and cardboard.

Procedure

1. *Construct the pentagon.* As in Activity 7, find the point that is approximately the center of the square. Using this point as one end, draw a 3 1/4-inch line perpendicular to a side of the square (fig. 1). With your compass, draw a circle using this line as a radius. Next, with your protractor, draw five adjacent 72-degree angles with vertices at the center. This can be done by first placing the protractor along the radius on one side (fig. 2) and then placing it along the radius on the other side. With a ruler, extend the sides of the angles to the circle. Join the ends of the radii to form a regular pentagon.

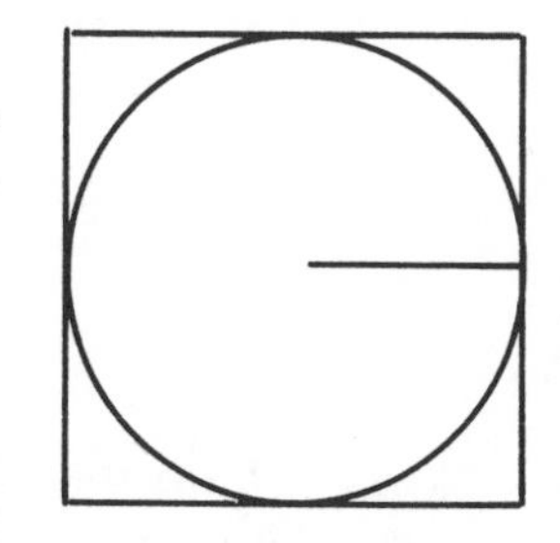

Fig. 1

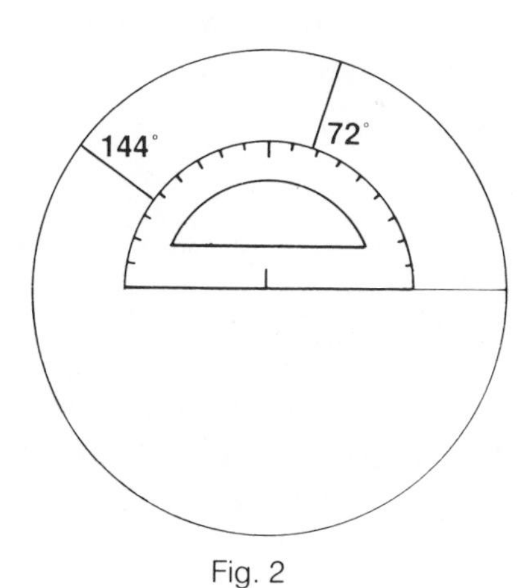

Fig. 2

2. *Mark the sides.* If only one side of the cardboard is colored, use that as the top and mark the uncolored side. With a pencil and ruler mark each side of the pentagon. Figure 3 shows the markings on one side. The measurements are as follows:

a) Mark the midpoint (approximately 1 7/8 inches from each end).

b) On each side of the midpoint measure off 1/8 inch.

c) From each of these points measure 1/4 inch.

d) From each of these points measure 1/2 inch.

e) The distance from the last mark to the vertex is approximately 1 inch.

That is, begin at the midpoint with a distance of 1/8 inch between marks, and then double the distance each time between succeeding marks.

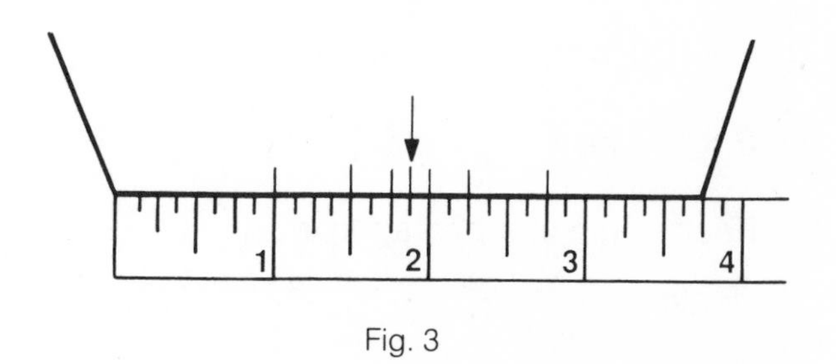

Fig. 3

3. *Cut out the pentagon and cut the slots.* Cut out the pentagon. Then cut a slot at least 1/8 inch deep on each of the marks (fig. 4).

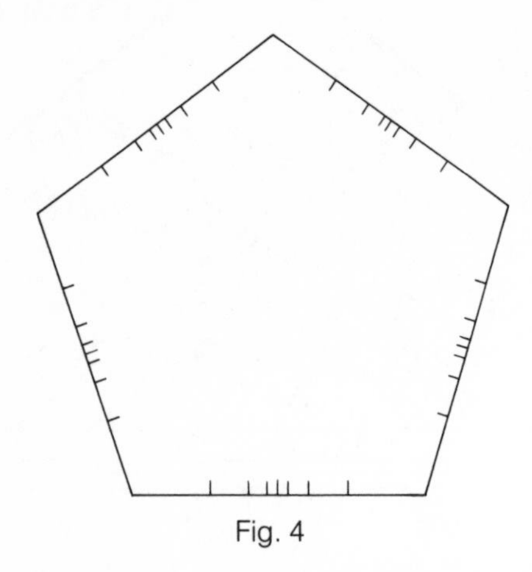

Fig. 4

4. *Prepare the string.* For variegated string, find the middle of the darkest part as indicated in figure 5, and fold at that point. Then measure a length of 2 feet 9 inches on the folded string (5 1/2 feet total length). Use this method to cut five pieces of string. To avoid tangling, wrap the pieces of cut string on a small rolled piece of paper. Put aside any leftover scraps of string.

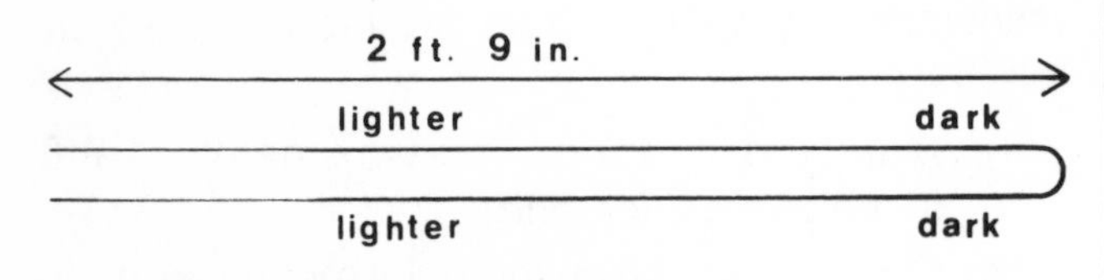

Fig. 5

5. *Weave the string.* Tie a knot at the end of one of the strings. Then, using the colored unmarked side of the pentagon as the top, pull the string up through slot 1 so that the knot is underneath. Follow the numbering in figure 6 by pulling the string over the pentagon to slot 2, under to 3, over to 4, under to 5, and so forth, ending with the string above from 13 to 14. Pull the string deep enough into each slot to allow a second string to fit into the same slot. The completed configuration will have two strings in each slot. Fasten the end of the string underneath with a small piece of tape.

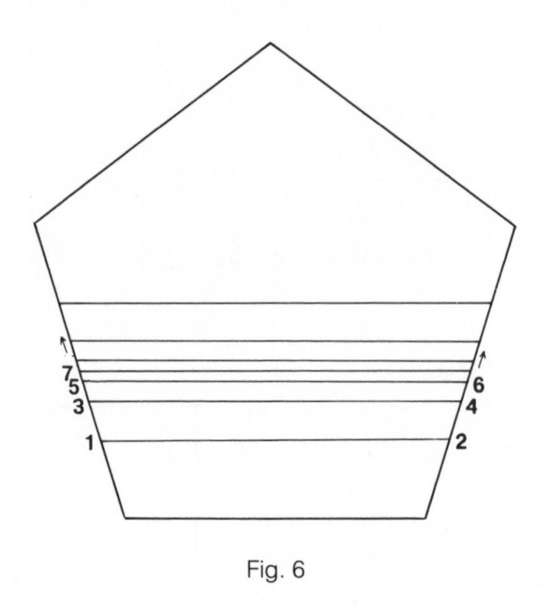

Fig. 6

Now perform this same weaving procedure for the other four sides of the pentagon. Each of the five strings should produce a set of lines parallel to one of the five sides.

When the design is finished, the darkest threads form a large star inscribed in the pentagon, and the lightest threads form a tiny star at the center of the pentagon.

Related Exercises

1. 1/2 of 5 ft. 6 in. = ______ ft. ______ in.

2. 2 × 1/8 in. = ______ in.

3. 2 × 1/4 in. = ______ in.

4. Each angle of a pentagon is an ____________ angle.

5. The five-sided government building in Washington, D.C., is called the ____________________ .

6. Draw all the lines of symmetry for a regular pentagon using the pentagon to the right.

7. A polygon with all its sides equal in length and all its angles equal in measure is called a ____________ polygon.

Teacher Notes for Activity 8 on page 26

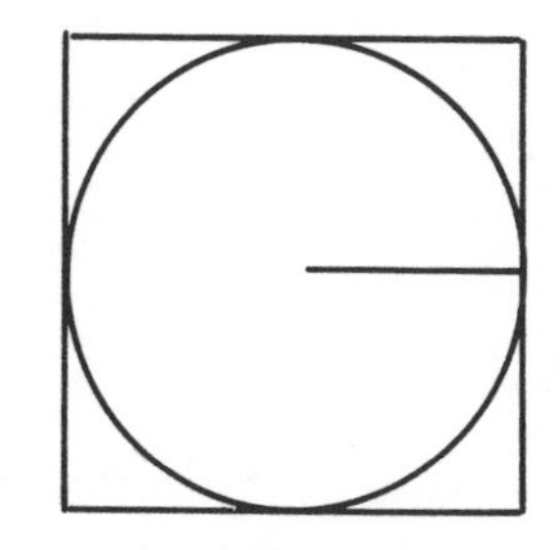

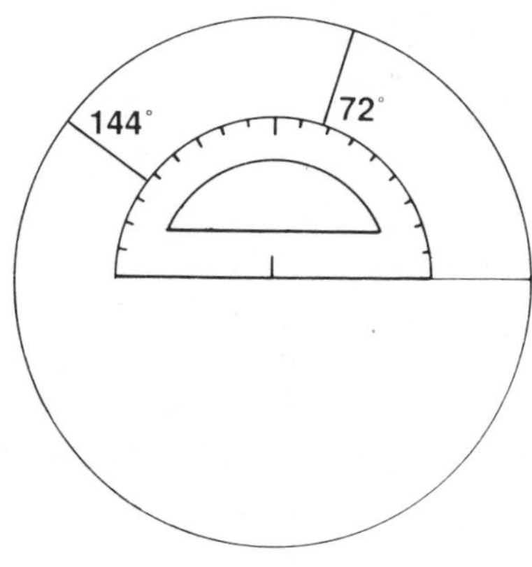

*ACTIVITY C

On Your Own

Circles in a Pentagon and Triangle

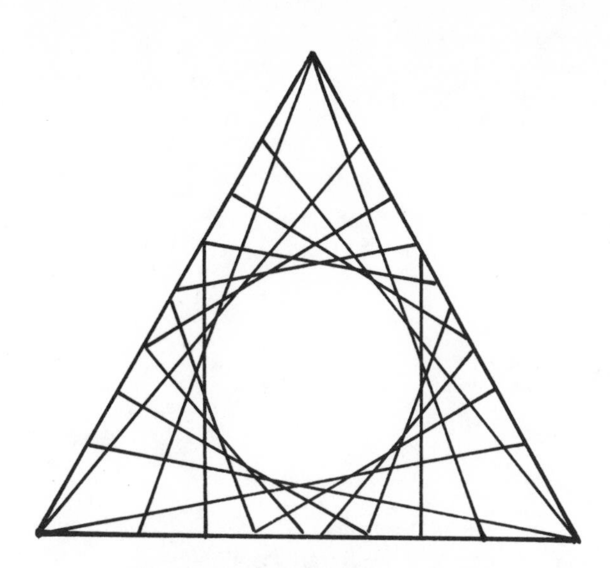

Based on Activities 1 and 6

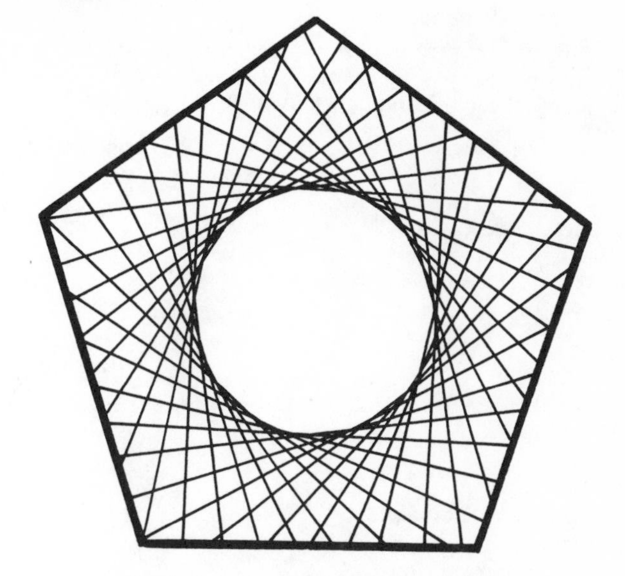

Based on Activities 6 and 7

With a procedure similar to that used in Activity 6, it is possible to construct a circle in any regular polygon. See if you can weave a circle in a triangle or pentagon.

TEACHER NOTES

*Activity C

Grades 8–10

See Activity 1 or 2 for triangle, Activity 7 for pentagon, and Activity 6 for string design.

TEACHER NOTES

Activity 5

Grades 4–7

Geometric concepts
Square
Curve
Identical curve (congruent curve)

Skills
Drawing marks 1/2 cm apart
Measuring 3.6 m and 2.7 m of string
Approximating 1/2 cm and 3 cm

Answers to Related Exercises

1. 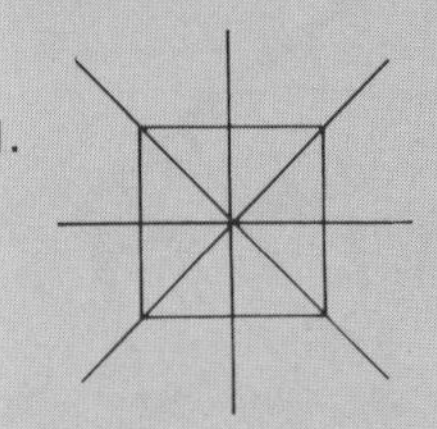
2. square
3. perpendicular, or congruent
4. right, or 90°

TEACHER NOTES

*Activity 6

Grades 8–10

Geometric concepts
Square
Circumscribed circle
Opposite vertices
Chord of circle
Radius
Arc and congruent arcs
10° angle
Central angle
Center of circle

Skills
Using compass to draw circle
Using protractor to draw 10° adjacent angles
Measuring 2 3/4 in. between tips of compass

Answers to Related Exercises

1. compass
2. protractor
3. radius
4. congruent
5. chord
6. congruent
7. circle

TEACHER NOTES

*Activity 8

Grades 7–10

Geometric concepts
Pentagon
Parallel lines
Adjacent angles
Vertices
Relationship between 1/8, 1/4, 1/2, and 1 in.
Equal central angles and equal chords
Measure of one complete revolution
Radius and radii

Skills
Drawing a 3 1/4-in. segment
Using compass to draw circle
Drawing 72° adjacent angles with protractor
Measuring 1/8, 1/4, 1/2, and 1 7/8 in.
Using ruler and protractor for precise measurements
Approximating 1/8 in.
Measuring 5 ft. 6 in. or 2 ft. 9 in. of string
Optional: finding and using the darkest part of variegated string

Answers to Related Exercises

1. 2 ft. 9 in.
2. 1/4
3. 1/2
4. obtuse, or 108°
5. pentagon
6. 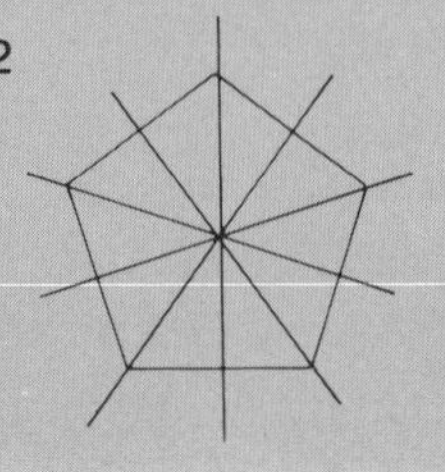
7. regular

ACTIVITY 9

Triangles in a Hexagon

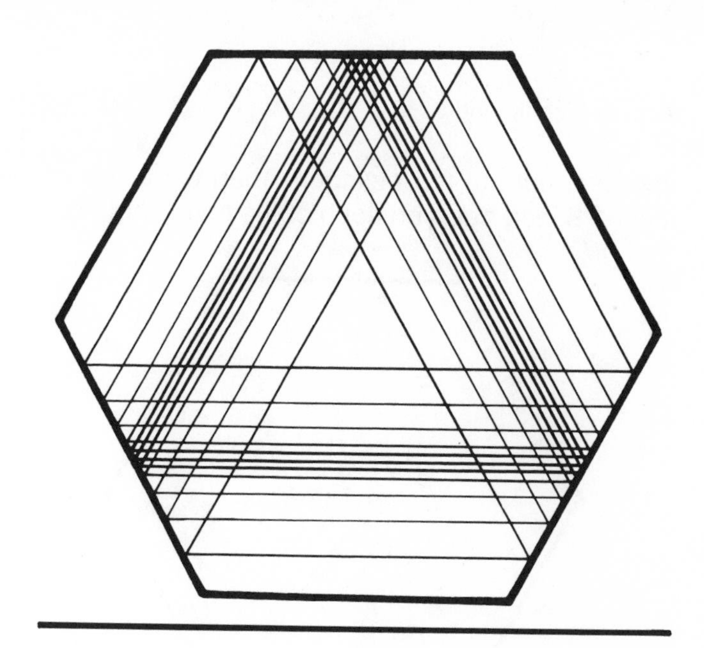

Materials

Protractor
Ruler
Pencil
Transparent tape, small amount
Scissors
Heavy cardboard, 9 in. square
String (preferably variegated), 31 to 40 ft.

Note: Use contrasting colors for string and cardboard.

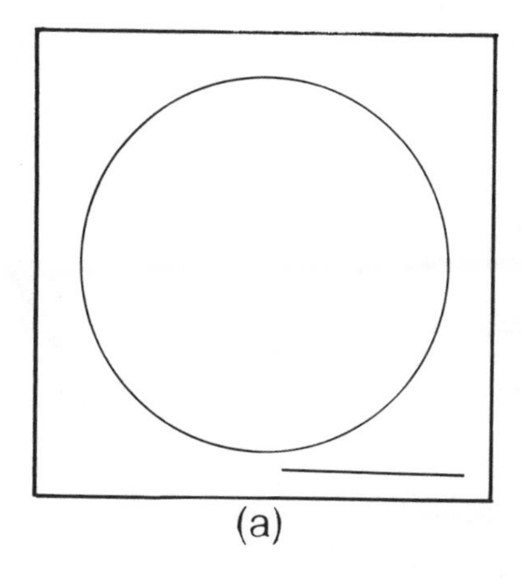

(a)

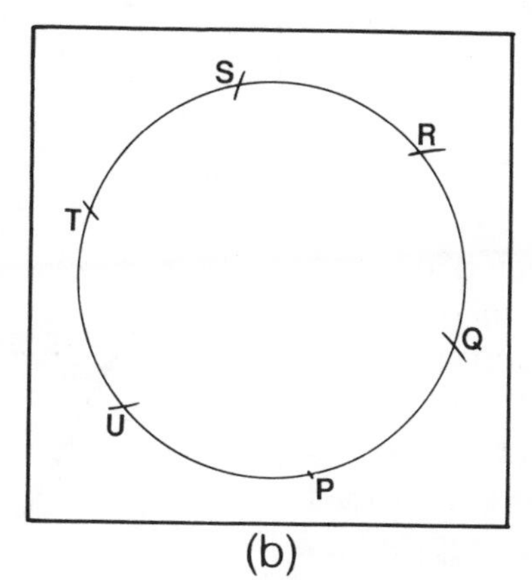

(b)

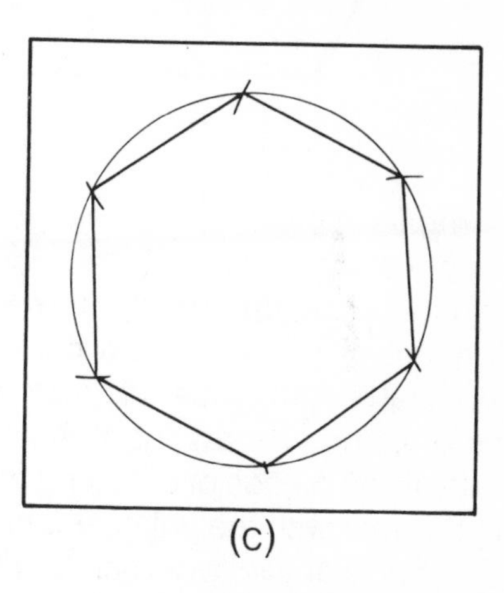

(c)

Fig. 1

Procedure

1. *Construct the hexagon.* Very near the bottom of your square, draw a line 4 inches long parallel to the edge. Setting your compass with this length as radius, draw a circle on the cardboard with your compass (fig. 1a). Next, keeping the 4-inch distance between tips of your compass, place the steel tip at any point, *P*, on the circle and draw an arc intersecting the circle at some point *Q*. Then place the steel tip on *Q* and draw an arc intersecting the circle at some point *R*. Continue this process to obtain points *S, T, U,* and *P* again. You should end where you began (fig. 1b). Draw line segments joining the points (fig. 1c). This six-sided figure is called a hexagon.

2. *Mark the sides.* Figure 2 shows how to mark one side of the hexagon. Use this figure as a guide to mark alternate sides of the hexagon. The measurements are as follows:

a) Mark the midpoint (2 inches from each end).

b) On each side of the midpoint measure off 1/8 inch.

c) From each of these points measure another 1/8 inch, then 1/4 inch, then 3/8 inch, then 1/2 inch.

d) The distance from the last mark to the end is 5/8 inch.

That is, begin at the midpoint with two lengths of 1/8 inch between marks and then add 1/8 inch to the length for each succeeding mark.

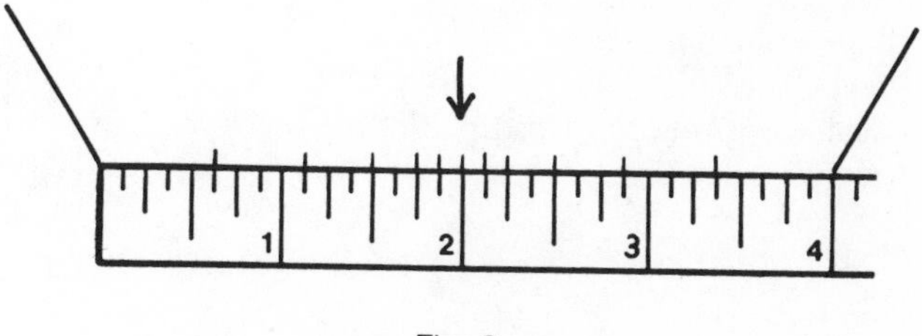

Fig. 2

3. *Cut out the hexagon and cut the slots.* With

scissors, cut out the hexagon. Then cut a slot at least 1/8 inch deep on each of the marks (fig. 3).

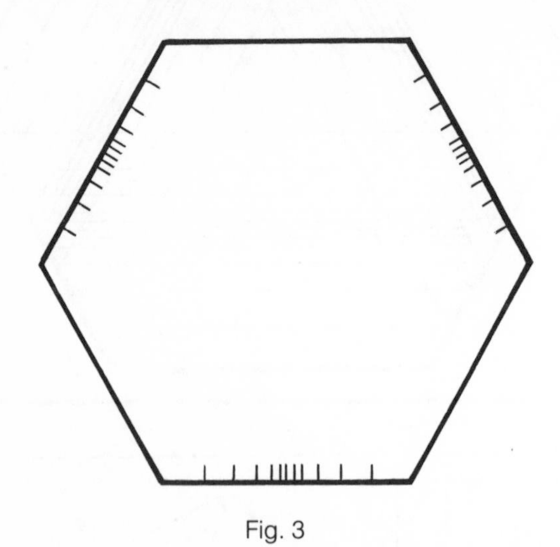

Fig. 3

4. *Prepare the string.* For variegated string, find the middle of the darkest part as indicated in figure 4. Then measure lengths of 5 feet 8 inches and 4 feet 9 inches on the two parts of the folded string (10 ft. 5 in. total length). Use this method to cut three pieces of string. Tie a knot at the end marked K. Wrap the three pieces of cut string on a small rolled piece of paper with the unknotted end wrapped on first. Put aside any leftover scraps of string.

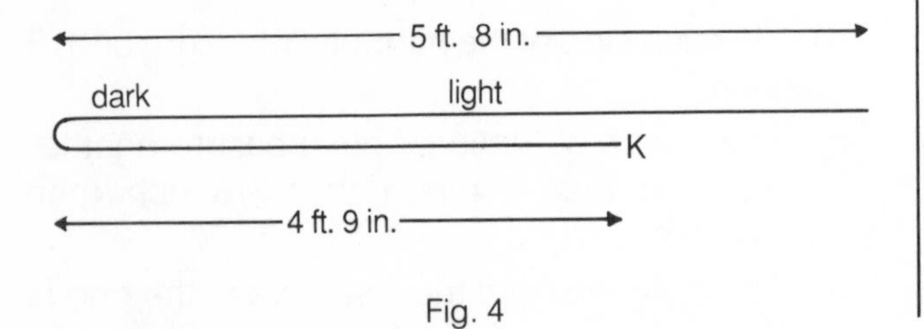

Fig. 4

5. *Weave the string.* To avoid tangling, unwind the string as you need it. Using the colored unmarked side as the top, pull the first string up through slot 1 so the knot is underneath. Follow the numbering in figure 5 by pulling the string across the top of the hexagon to slot 2, underneath to 3, over to 4, back to 5, and so forth. Pull the string deep enough into each slot to allow a second string to fit into the same slot. The completed configuration will have two strings in each slot. Fasten the end of the string underneath with a small piece of tape.

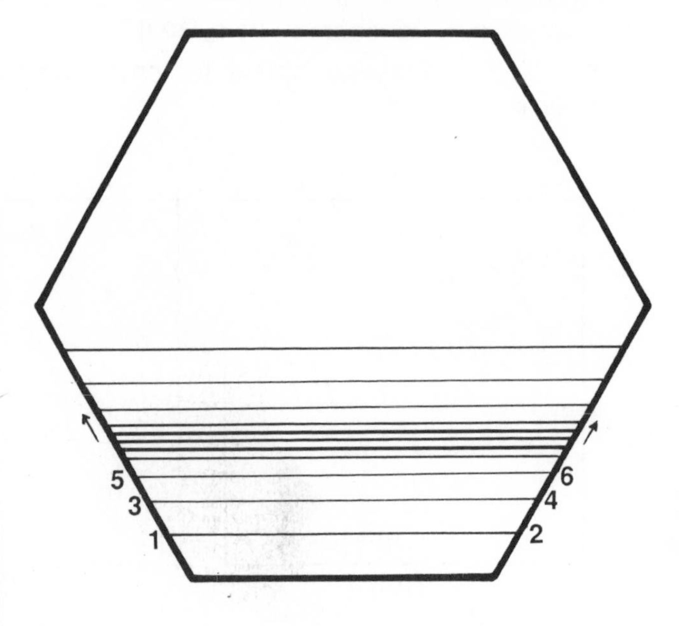

Fig. 5

Now perform this same weaving procedure for the other marked sides of the hexagon. Each of the three strings should produce a set of lines parallel to one of the three unmarked sides. Always begin with the knotted end of the string so that the darkest part joins midpoints of sides.

TEACHER NOTES

Activity 9

Grades 5–9

Geometric concepts
Parallel lines
Relationship between 1/8, 1/4, 3/8, 1/2, and 5/8
Alternate sides
Hexagon
Circle
Arcs
Radius
Line segment
Naming points with capital letters

Skills
Drawing 4-in. segment
Measuring 1/8, 1/4, 3/8, 1/2, and 5/8 in.
Measuring 4 ft. 9 in. and 5 ft. 8 in. of string
Using ruler for precise measure
Constructing a circle with radius 4 in.
Drawing arcs with compass
Optional: finding and using the darkest part of variegated string

ACTIVITY D

On Your Own

Star in a Hexagon

Based on Activities 4 or 7 and 9

After studying Activities 4 and 7, you might like to try this six-pointed star in a hexagon.

TEACHER NOTES

Activity D

Grades 5–9

See Activity 9 for hexagon and Activity 4 or 7 for string design.

ACTIVITY E

On Your Own

Parabolic Curves

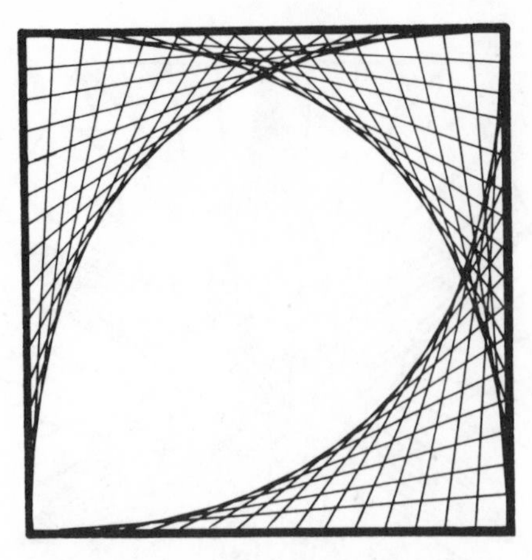

Based on Activities 4 and 5

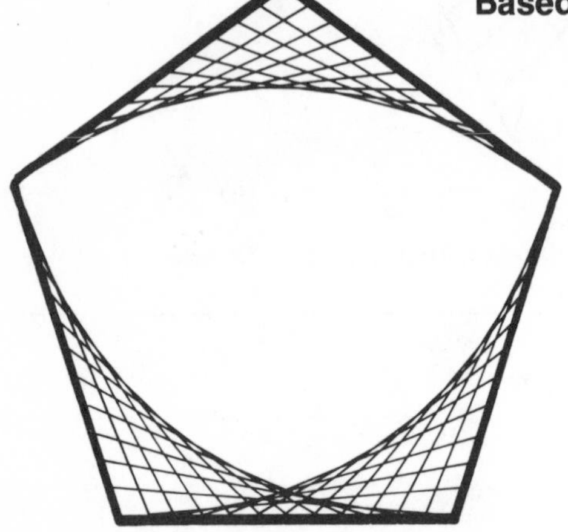

Based on Activities 5 and 7

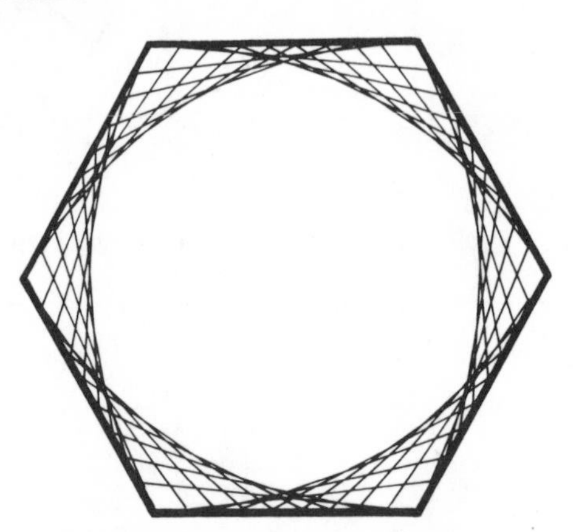

Based on Activities 2 and 9

Curves like those described in Activities 2 and 5 are called parabolas. A parabola can be woven in any angle. The designs above show several parabolas in each polygon. Weave one of these, or create your own pattern with parabolas in a polygon.

TEACHER NOTES

Activity E

Grades 4–9
E-1: Grades 4–6
E-2: Grades 6–9
E-3: Grades 5–9

See Activity 4 for square, Activity 7 for pentagon, Activity 9 for hexagon, and Activity 2 or 5 for string design.

ACTIVITY F **On Your Own**

Designs Using Variegated String

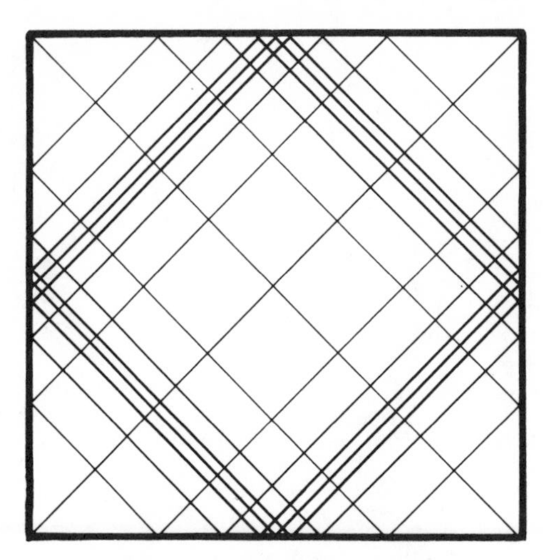

Based on Activities 4 and 9

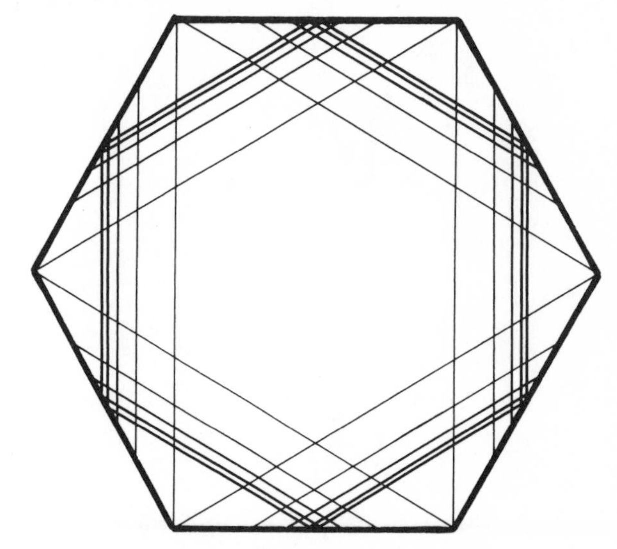

Based on Activities 8 and 9

These two designs use the same measuring techniques as those used in Activities 8 and 9. The challenge is to get the darkest part of the string to form a square in the first design and a hexagon in the second design.

TEACHER NOTES

Activity F

Grades 6–10

See Activity 4 for square, Activity 9 for hexagon, and Activity 8 or 9 for string design.

ACTIVITY 10

Stellated Hexagon in a Hexagon

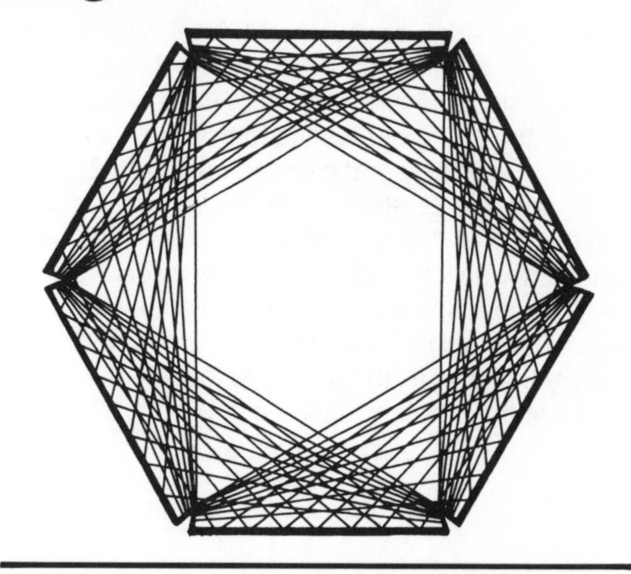

Materials

String, 17.8 m
Compass
Centimeter ruler
Pencil
Scissors
Tape
Stiff cardboard, 20 cm square

Procedure

1. *Construct the hexagon.* Near the bottom of your square, draw a line 8 cm long. If you do not have a centimeter ruler, cut out and use the one on page 35. Setting your compass with 8 cm as radius, draw a circle on the cardboard with your compass (fig. 1a). Next, keeping the 8-cm distance between tips of your compass, place the steel tip at any point *P* on the circle and draw an arc intersecting the circle at some point *Q*. Then place the steel tip on *Q* and draw an arc intersecting the circle at some point *R*. Continue this process to obtain points *S, T, U,* and then *P* again. You should end where you began (fig. 1b). Draw line segments joining the points (fig. 1c). This six-sided figure is called a hexagon. Cut out the hexagon.

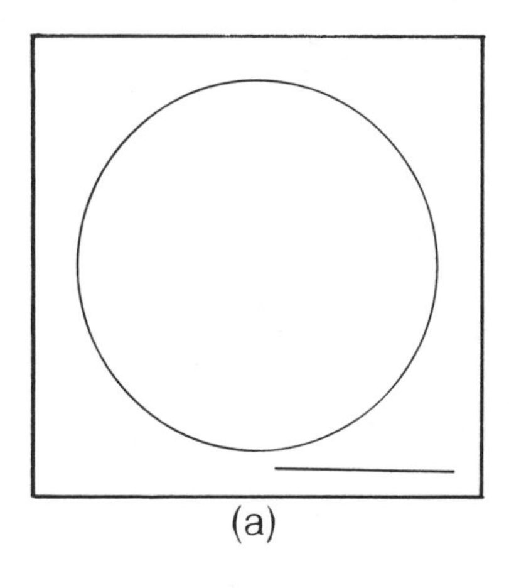

(a)

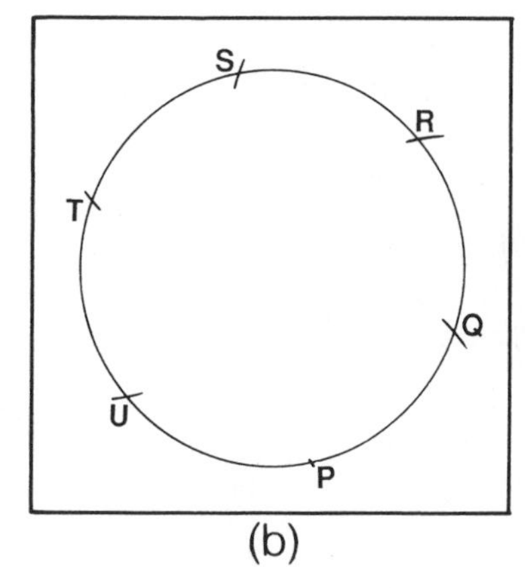

(b)

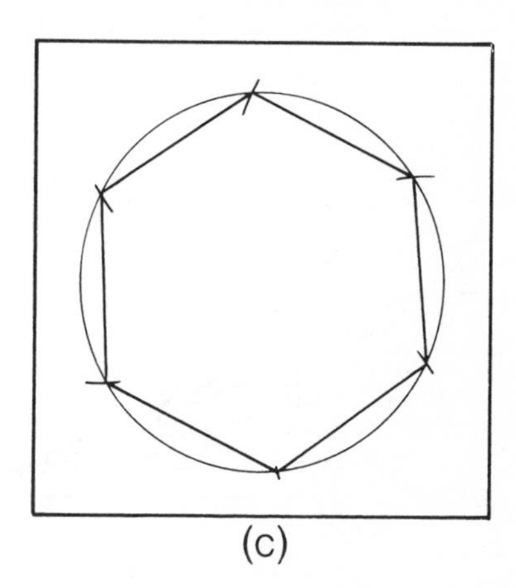

(c)

Fig. 1

2. *Mark and cut the sides.* Mark every 1 cm on each side of the hexagon (fig. 2). Cut a slot about 1/2 cm deep on each mark.

3. *Cut the vertices.* At each vertex cut a wedge at least 1/2 cm deep. Then, on the underside of the cardboard, label the vertices *P, Q, R, S, T,* and *U* (fig. 2).

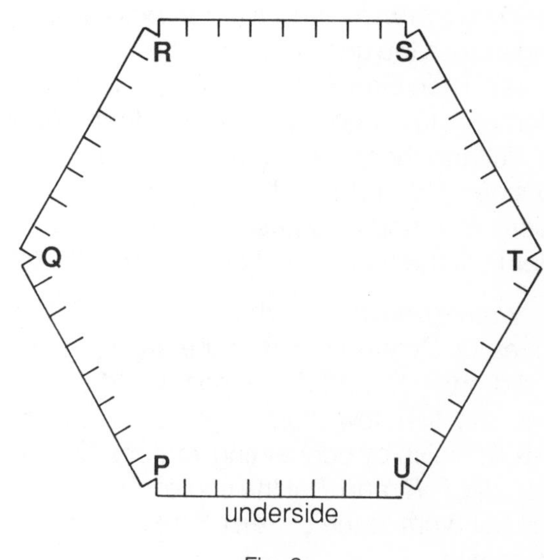

Fig. 2

4. *Weave from the first vertex.* Keep the string on a spool or rolled paper to prevent tangling, then unwind the string as you need it. Knot the end of the string. Now follow figure 3. Bring the string up through slot 1 securing the knot underneath. Then stretch the string across the top of the hexagon to vertex *P*, then across the underside to slot 2, above to *P*, under to 3, above to *P*, and so on, until you have used all the slots on side *TU*. Pull the string deep enough into each slot to allow a second string to fit into the same slot. Next weave the same pattern to the right of *P* using slots 1′, 2′, 3′, and so on. Figure 3 shows all the strings to the left of vertex *P* and just the first few to the right.

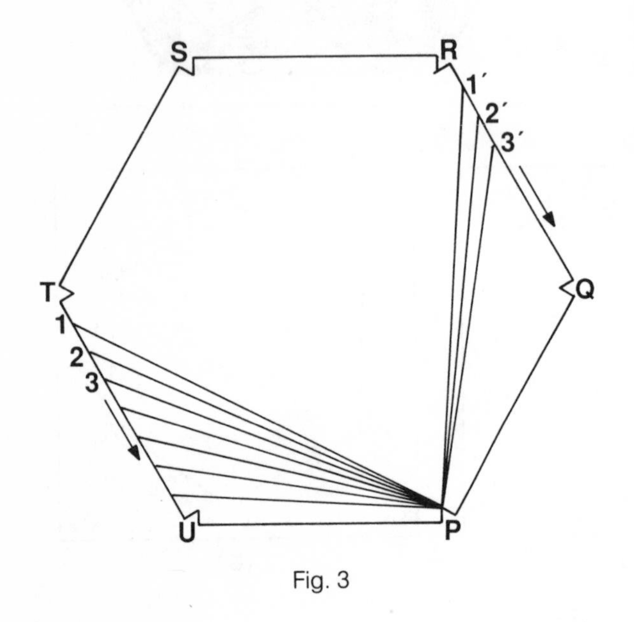

Fig. 3

5. *Weave from the next two vertices.* Now pull the string across the underside of the hexagon to slot 1 (fig. 4a). Then stretch it across the top to vertex *R*, underneath to 2, above to *R*, and so forth. Complete side *PQ* and then side *TS*. Next weave from vertex *T* to sides *RS* and *PU* (fig. 4b). Now connect the vertices *R*, *T*, and *P* by passing your string over and under until the triangle in figure 4c is complete.

6. *Weave from the last three vertices.* For each of vertices *Q*, *S*, and *U*, weave the same design that you did on *P*, *R*, and *T* in steps 4 and 5. Figure 5 shows the first few slots used for weaving from vertex *S*. Finish by connecting vertices *Q*, *S*, and *U* as you did *P*, *R*, and *T* at the end of step 5. Then cut off all but 3 cm of the extra string and tape the end underneath.

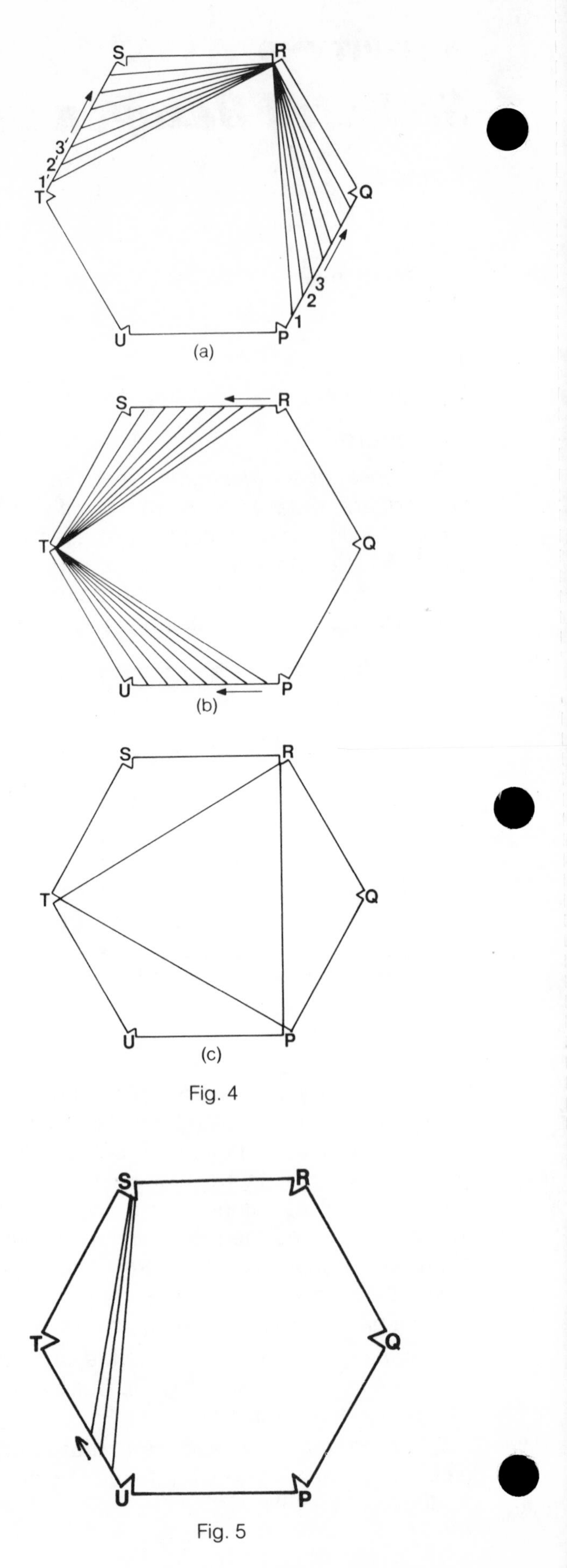

Fig. 4

Fig. 5

Related Exercises

1. A polygon having six sides is called a ____________ .
2. Each angle of a regular hexagon is a (an) ____________ angle.
3. The opposite sides of a regular hexagon are ____________ to each other.
4. Each side of a regular hexagon has the same length as the ____________ of its circumscribed circle.
5. A circle circumscribed around a regular polygon passes through each ____________ of the polygon.
6. Draw all the lines of symmetry for a regular hexagon using the hexagon below.

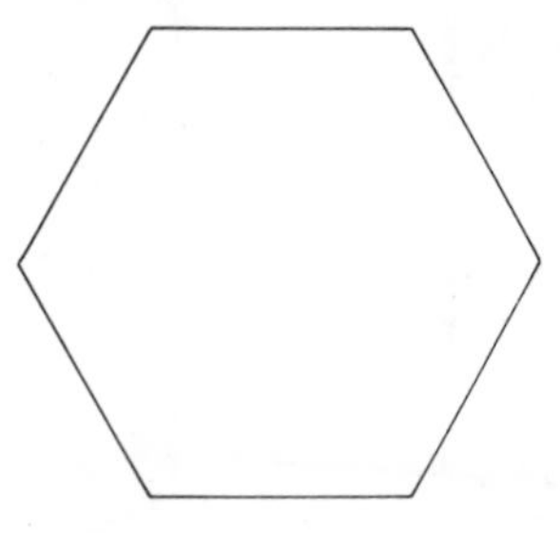

7. In the regular hexagon, ______ lines of symmetry pass through opposite vertices and ______ lines of symmetry pass through midpoints of opposite sides.
8. The number of straight lines that can be drawn through both of two distinct points *P* and *Q* is ______ .
9. A side of a polygon with endpoints *R* and *S* can be called side ______ or side ______ .

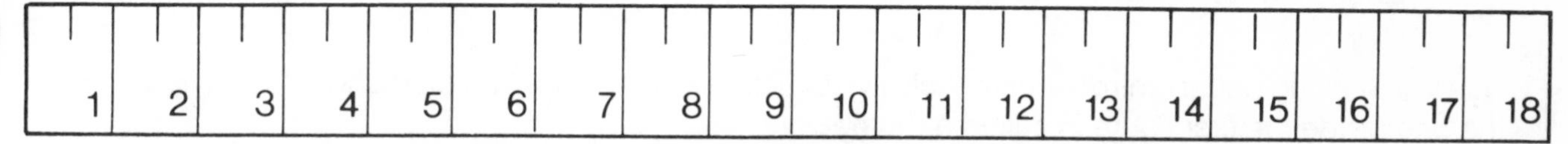

Centimeter ruler

TEACHER NOTES

Activity 10

Grades 6–9

Geometric concepts
Hexagon
Circle
Arcs
Radius
Triangle
Vertex and vertices of polygon
Naming points with capital letters
Naming vertices with capital letters
Naming sides of a polygon with two letters

Skills
Using compass to draw circle
Drawing arcs with compass
Measuring 1/2 cm with ruler
Approximating 1/2 cm and 3 cm

Answers to Related Exercises

1. Hexagon
2. Obtuse, or 120°
3. Parallel, or congruent
4. Radius
5. Vertex
6.
7. 3; 3
8. 2
9. *RS; SR*

ACTIVITY G **On Your Own**

Rhombus and Triangle in a Hexagon

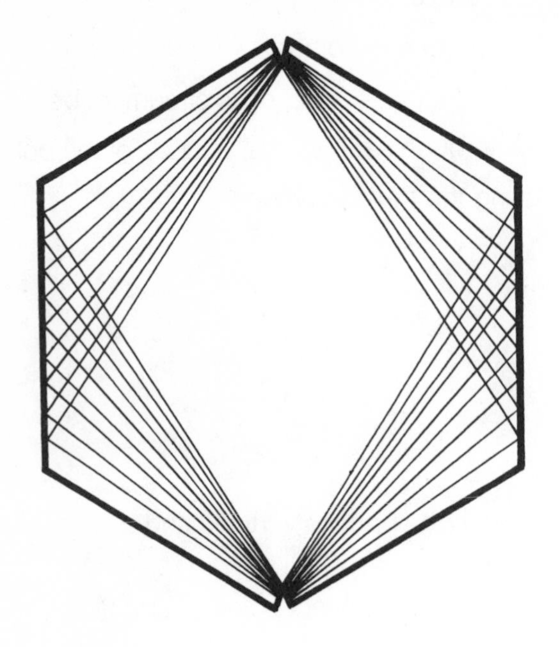

Based on Activities 9 and 10

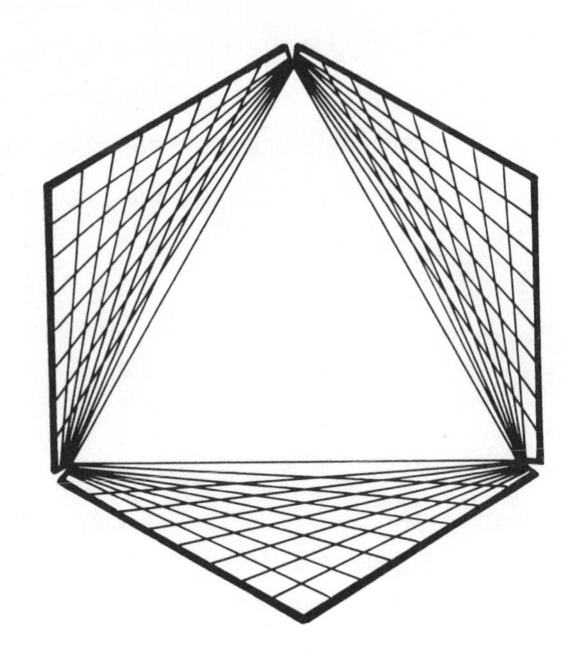

Based on Activities 9 and 10

Use the weaving pattern introduced in Activity 10 for these two designs. A diamond-shaped figure like the one inside the first figure is called a rhombus.

TEACHER NOTES

Activity G

Grades 6–9

See Activity 9 for hexagon and Activity 10 for string design.

ACTIVITY 11

Curved Surface in a Tetrahedron

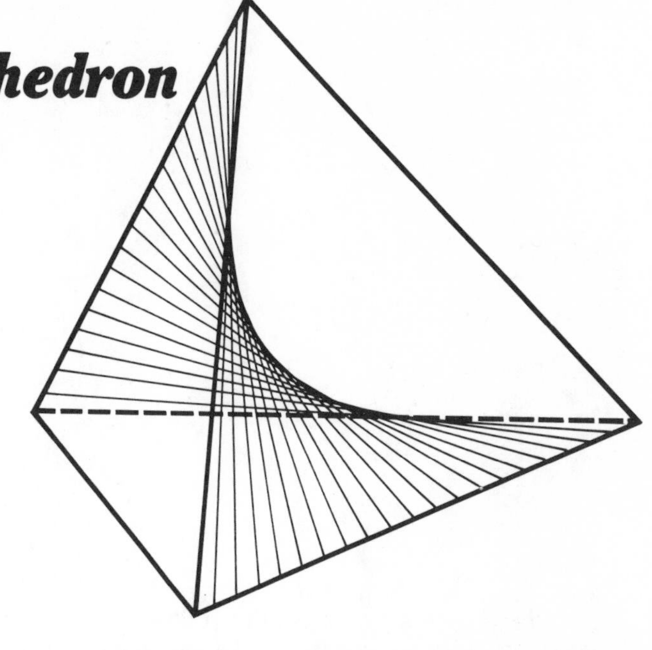

Materials

Ruler
Glue
Thin string, 13 ft.
6 plastic straws
2 needles, one heavy and one rather thin

Procedure

1. *Prepare the straws.* Cut the straws to a length of 5 inches. Punch holes through two of the straws at each 1/4 inch with any needle or pin. To punch the holes, lay the straw down and punch all the way through so you have two rows of holes in a straight line. Be sure to hold the needle or pin vertically so the holes on the bottom of the straw will also be spaced equally and form a straight line.

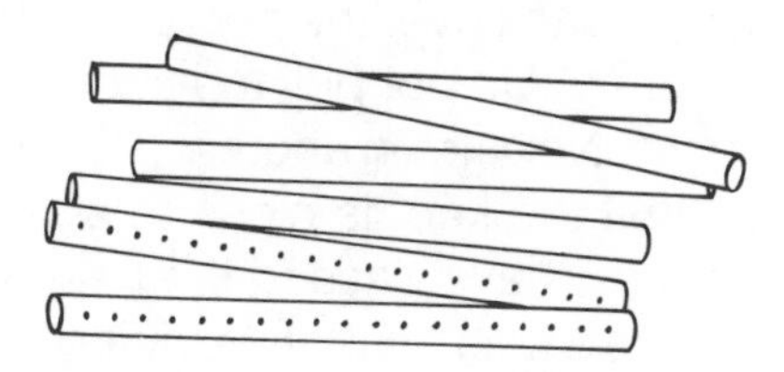

Fig. 1

2. *Build the tetrahedron.* Thread the heavy needle with 4 feet of thin string. Drop the threaded needle through the interior of three straws, the second straw being one that has punched holes. Next tie the string to form a triangle with no slack in the string (fig. 2a). Now drop the threaded needle through the second punched straw and then through one more straw (fig. 2b). Then drop the needle back through the first punched straw and then through the last of the six straws (fig. 2c). Pull the string tight. Finally drop the needle through the second punched straw to return to the starting point (fig. 2d). Tie the string securely and cut off the ends.

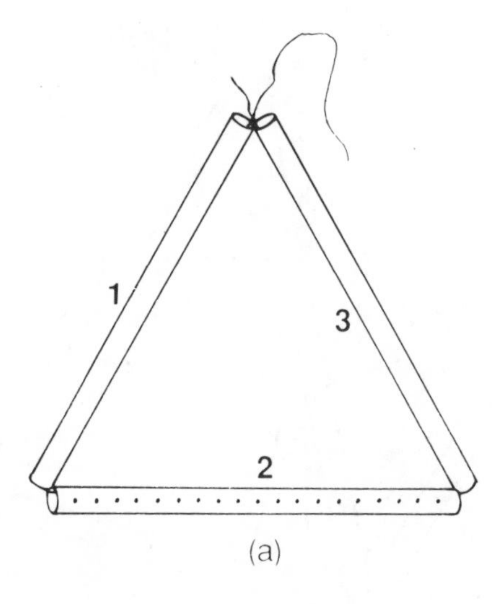

(a)

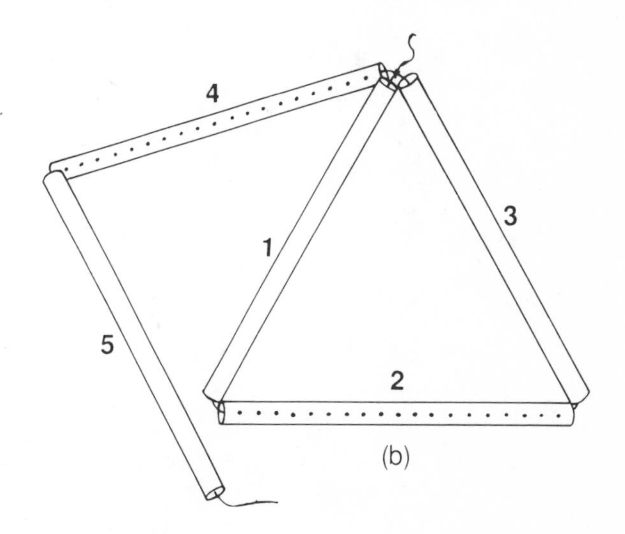

(b)

Fig. 2

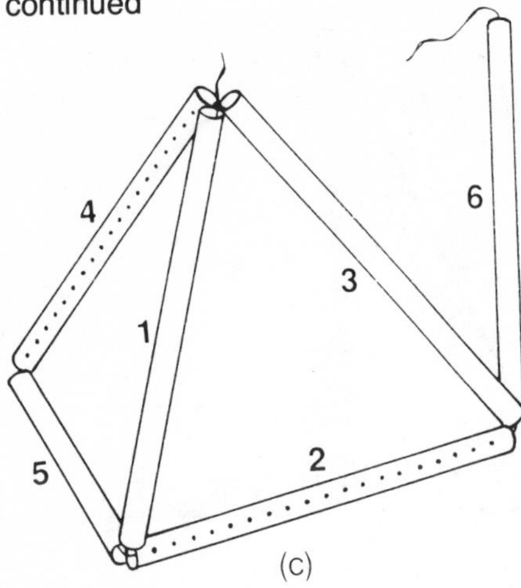

(c)

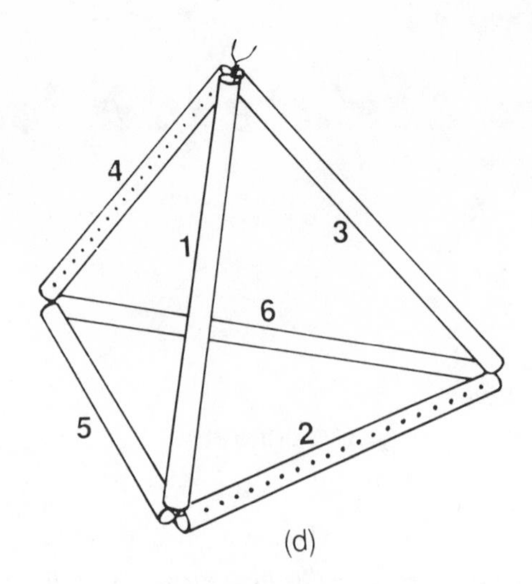

(d)

Fig. 2 (continued)

3. *Stabilize the vertices.* Put a little glue on the inside of each of the four vertices so the three straws meeting at each vertex are immovable. Position the punched straws so that each has a line of holes facing the interior of the tetrahedron. If necessary push the straws a little so that each vertex is arranged neatly. Allow the glow to dry, preferably overnight.

4. *Weave the string.* Thread the smaller needle with approximately 3 feet of string and knot the end of the string so that it will not slip through a hole. Now follow the numbering in figure 3. (Fig. 3 indicates only the first five holes on each of the straws to avoid cluttering the drawing.) Push the needle through hole 1, from exterior to interior. Following figure 4, weave the string in this manner: when the string connects a hole on a straw with the next hole on the same straw, it lies on the outer side of a straw; when the string connects the two straws, it lies inside the straw frame. Pass the needle through hole 2 and then through hole 3. Next stretch the string to holes 4 and 5 on the first straw, then on to 6, 7, 8, and so on, as shown in figure 4. When you are out of string, go back over the design, taking up any slack in the string. Do not pull too tightly or you will arch the straws. Next tie on another 3 feet of string in such a way that the knot is close to a straw. Then proceed until you have passed the needle through the last hole. Tie a knot so that the string cannot slip through the hole, and cut off the remaining string. Glue each knot against a straw to make the knots less noticeable. You have created a curved surface in a tetrahedron.

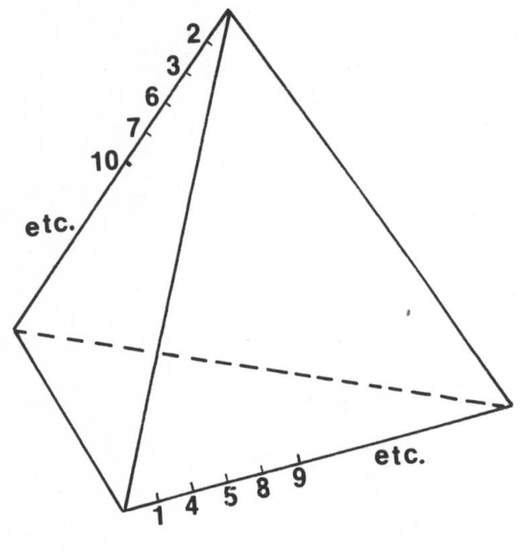

Fig. 3

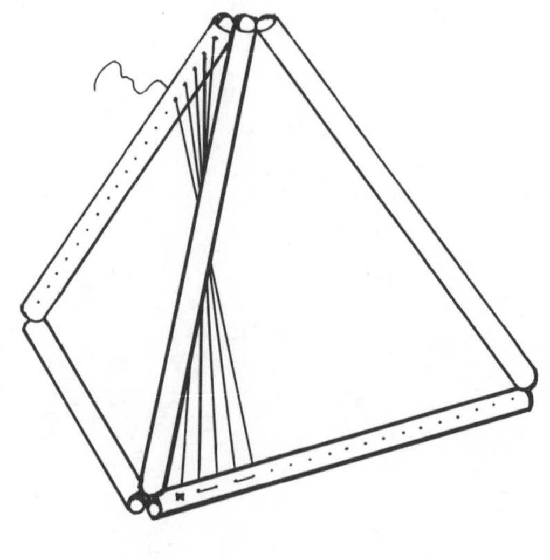
Fig. 4

Related Exercises

1. A tetrahedron has ______ edges, ______ faces, and ______ vertices.
2. The six edges of the regular tetrahedron form ______ pairs of skew lines.
3. ______ of the edges of a regular tetrahedron are parallel to each other.
4. A pyramid with a triangular base is called a (an) ________________ .
5. Weaving on skew lines produces curved surfaces. Skew lines are lines that do not ________________ and are not ________________ .

TEACHER NOTES

Activity 11

Grades 7–10

Geometric concepts
Tetrahedron
Vertex
Edge
Plane angle
Polyhedral angle
Pyramid
Skew lines
Vertical
Curved surface
Interior and exterior of tetrahedron

Skills
Measuring 1/4 in.
Measuring 3 or 4 ft. of string
Punching a straight row of holes through a straw
Threading a needle
Tying a knot
Weaving on two skew edges of a straw tetrahedron

Answers to Related Exercises

1. 6; 4; 4
2. 3
3. none
4. tetrahedron
5. intersect; parallel

ACTIVITY H

On Your Own

Two-Layer Curved Surface in a Tetrahedron

Based on Activity 11

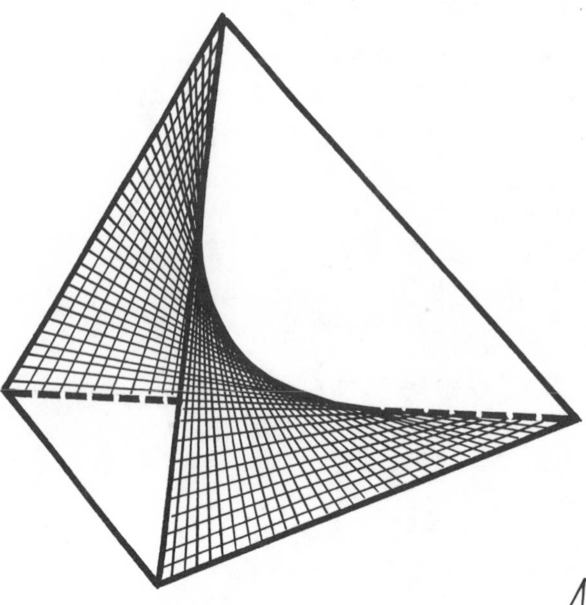

For this model you will need six straws, four of them punched. The punched straws are labeled a, b, c, and d in figure 1. After constructing the tetrahedron with the punched straws in the appropriate places, use step 4 of Activity 11 on straws a and d. Then use this same step on straws b and c, choosing the first holes in such a way that the second curved surface lies on the first curved surface. Use two colors, one for each layer.

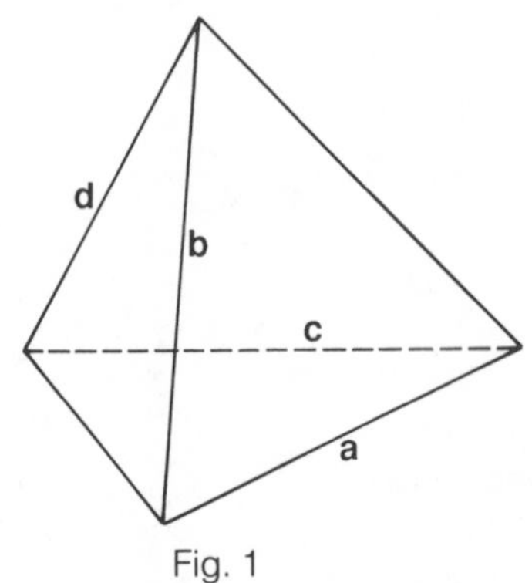

Fig. 1

TEACHER NOTES

Activity H

Grades 7–10

See Activity 11

*ACTIVITY 12

Two Curved Surfaces in a Tetrahedron

Materials

Ruler
Glue
2 blue strings, each 3 ft. long
2 pink strings, each 3 ft. long (Any two colors will do.)
Thin string, 5 ft.
6 plastic straws, each 4 in. long
2 needles, one heavy and one rather thin
Copy of Activity 11

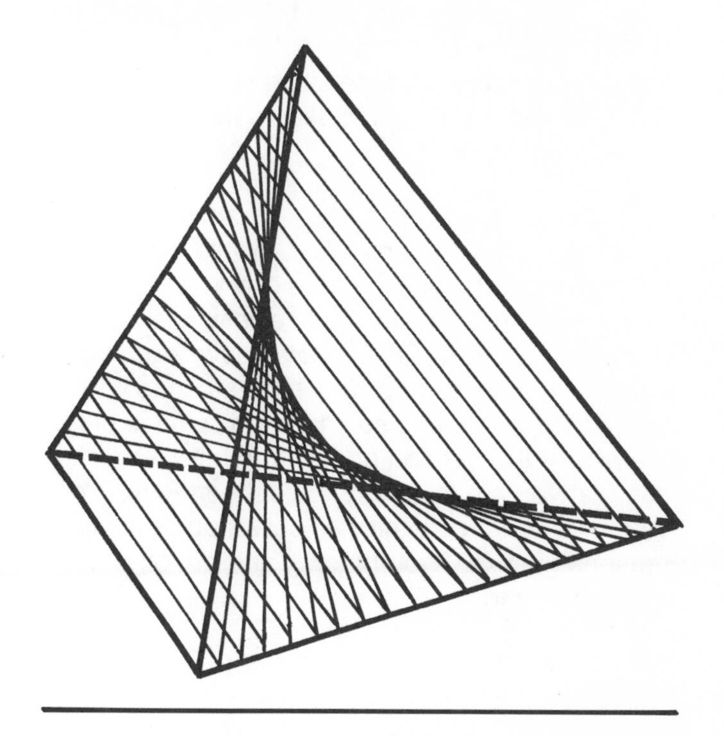

Procedure

1. *Prepare the straws.* Follow the directions in step 1 of Activity 11, but cut the straws 4 inches long because shorter straws can better withstand the tension from the many strings.

2. *Build the tetrahedron.* Follow the directions in step 2 of Activity 11.

3. *Stabilize the vertices.* Find each vertex where one of the straws is not connected to both its neighbors. For example, in figure 1, straw *b* is connected to straw *a* but not to straw *c*. To connect *b* to *c*, drop the threaded needle through *b* and then through *c*. Repeat this process until every vertex has the kind of connection shown in figure 2. That is, make sure each straw of each vertex is connected to both its neighbors. These extra strings will help the angles hold their shape during the weaving. Now glue the vertices as in step 3 of Activity 11.

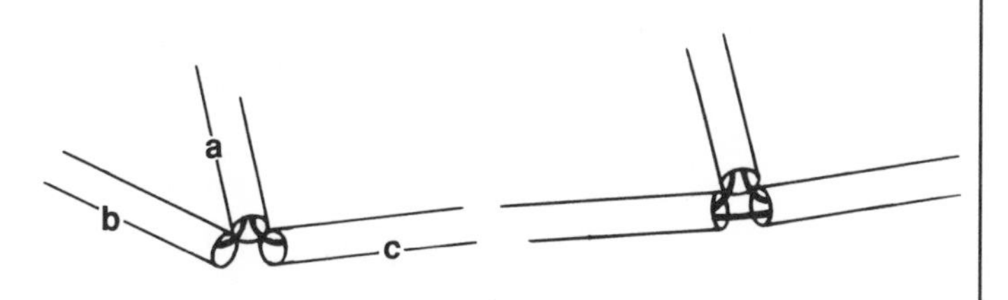

Fig. 1 Fig. 2

4. *Weave half of the first curve.* Thread the smaller needle with a piece of blue string and knot one end so that it will not slip through a hole. Now follow figure 3; that is, begin at the middle hole, h, which is the eighth one from each end. Then stretch the string through the interior of the figure and through hole w, the middle hole of the other punched straw. Next push the needle from outside the figure through hole x. Then pull the string back through hole i, then through hole j, and on to y, z, k, l, a′, and so on, until you have completed the half-curve shown in figure 3. If you have done the weaving correctly, the outside of each punched straw

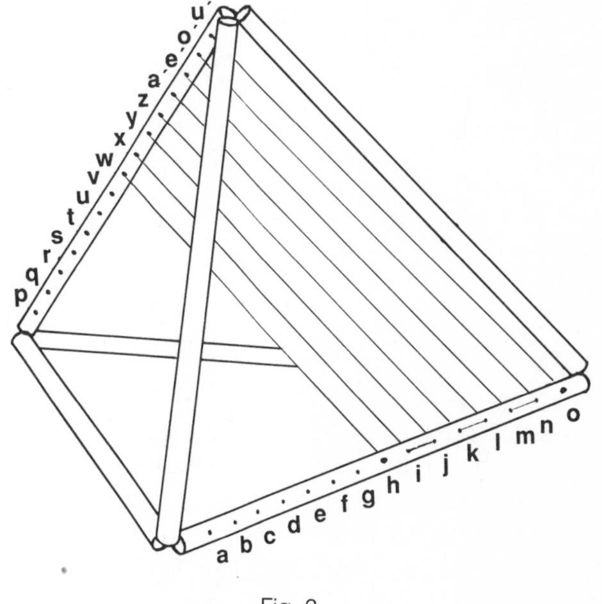

Fig. 3

contains the knots and a row of stitches, and the inside contains strings leading to the other straw. Figure 3 shows the knots-and-stitches side of the bottom straw and the strings leading downward from the top straw. Pull the string rather tightly so it does not sag. Tie a knot to secure the end and cut off the extra string.

5. *Weave half of the second curve.* Use pink string for this half-curve. This step follows the same procedure as step 4. Use holes h to o on the bottom straw again. Also use the middle hole, w, on the top straw again, but on this straw weave downward. That is, begin at h, then stretch the string to w, then on to v, i, j, u, t, k, and so on. After you have pulled the pink string through about three holes, check to see that you are following figure 4 correctly. When you have completed this half-curve, pull the thread reasonably tight, knot the end, and cut off the unneeded thread.

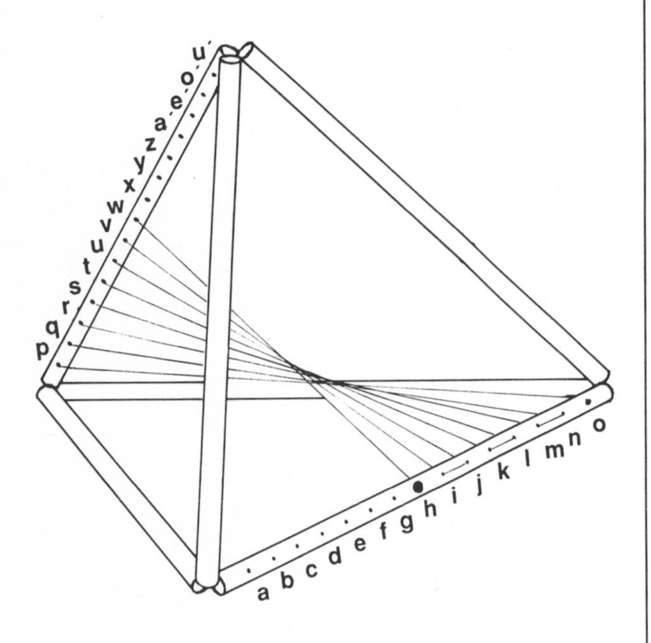

Fig. 4

6. *Complete the first curve.* Use blue string and the same basic procedure as in steps 4 and 5. Do not use the middle hole on either straw. Work with the other half of the bottom straw and the holes with pink string at the top. Following figure 5, begin at hole g, then stretch the string to v, to u, to f, to e, to t, and so on.

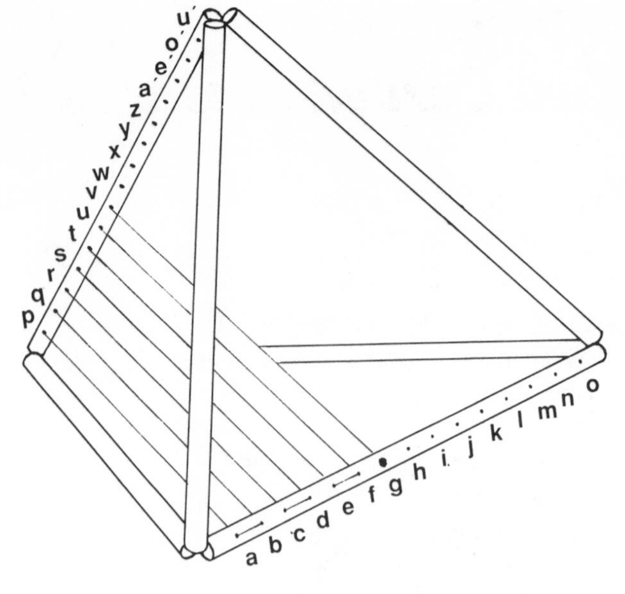

Fig. 5

7. *Complete the second curve.* Use pink string and the holes that contain only a blue string. When step 7 is completed, every hole should contain both a pink string and a blue string. Begin at hole g, then on to x, to y, to f, to e, to z, and so forth (fig. 6).

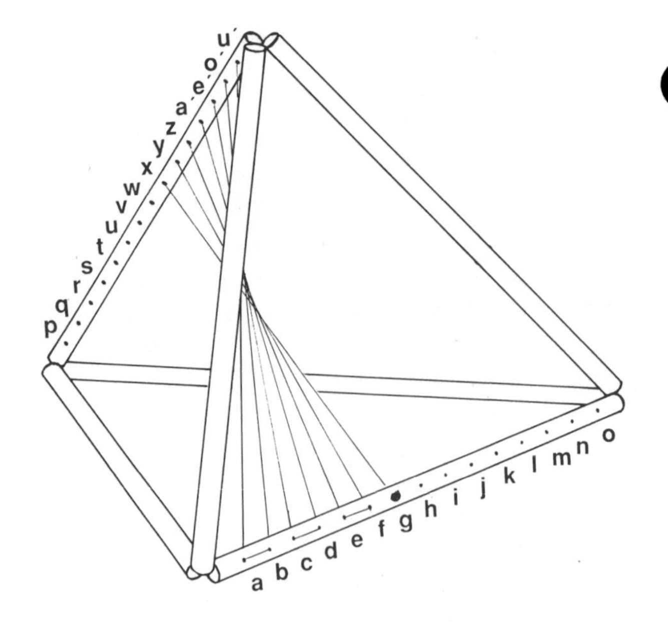

Fig. 6

The completed design consists of two congruent copies of the curved surface constructed in Activity 11. These two curved surfaces open in different directions and intersect in the middle, forming an attractive and very symmetrical design. An interesting feature of this configuration is that the view interior to each of the four faces is exactly the same.

Related Exercises

1. In a tetrahedron, _______ faces meet at each vertex.
2. A regular tetrahedron has _______ lines of symmetry that connect a vertex to a center of a face and _______ lines of symmetry that connect midpoints of a pair of skew edges.
3. In a regular tetrahedron all the ____________________ are congruent, all the ____________________ are congruent, and all the ____________________ are congruent.

TEACHER NOTES

*Activity 12

Grades 8 and above

Geometric concepts
Tetrahedron
Interior of tetrahedron
Exterior of tetrahedron
Midpoint of a segment (straw)
Congruent curved surfaces
Face of tetrahedron
Skew lines
Intersecting curved surfaces

Skills
Measuring 4 in. and 1/4 in. on a straw
Weaving on two skew edges of a straw tetrahedron
Punching a straight row of holes through a straw
Threading a needle
Tying a knot

Answers to Related Exercises

1. 3
2. 4; 3
3. Faces; edges; angles

ACTIVITY 13

Four Curves on a Tetrahedron

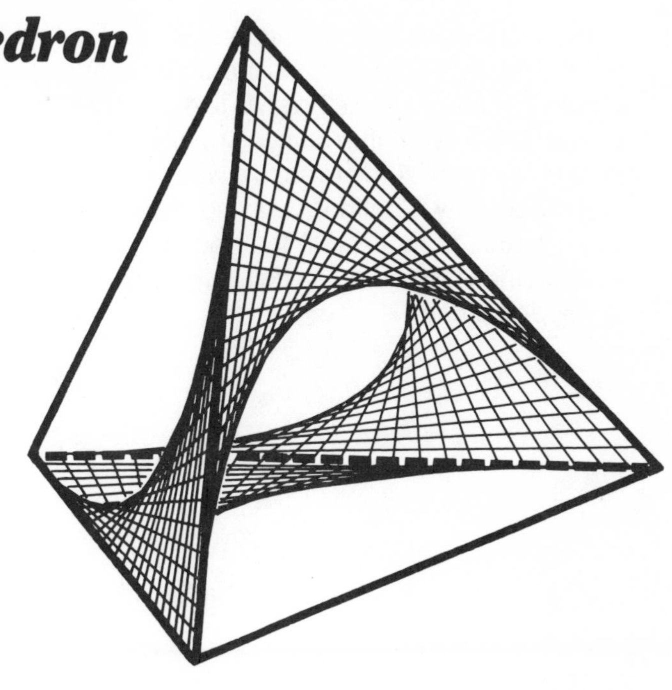

Materials

String, 10 m
Scissors
Glue
Metric ruler
6 plastic straws
2 needles, one heavy and one rather thin
Copy of Activity 11

Procedure

1. *Prepare the straws.* With scissors cut the six straws to a length of 12 centimeters. Then punch holes with a needle or pin through four of the straws at every 1/2 centimeter (fig. 1). The ruler below may be used for measuring the straws.

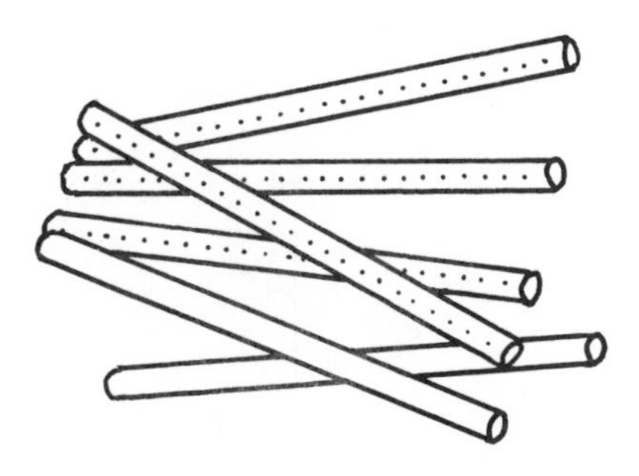

Fig. 1

2. *Build and stabilize the tetrahedron.* Follow the instructions in steps 2 and 3 of Activity 11, but interchange punched and unpunched straws. Also, when gluing the vertices, position each punched straw so that a smooth section between lines of holes faces the interior of the tetrahedron.

3. *Weave the curves.* Before beginning to weave the string, study the pattern inside the triangle (fig. 2). This design, a parabola, is to be woven in each of the four faces of the tetrahedron. Now thread the smaller needle with about a meter of string. Knot the end of the string so that it will not slip through a hole.

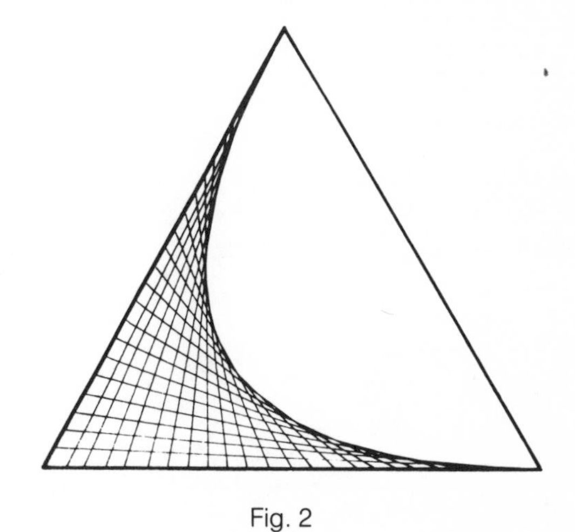

Fig. 2

1	2	3	4	5	6	7	8	9	10	11	12	13	14	15	16	17	18

Centimeter ruler

Then follow the numbering in figure 3. Push the needle through hole 1. Next stretch the string over to hole 2, then up to hole 3, down to 4, over to 5, on to 6, and so on. When the string is nearly gone, tie on another meter of string in such a way that the knot is close to a straw. Continue weaving and tying on the string in this manner until the needle has passed through the last hole. Then tie a knot and cut off the unneeded string.

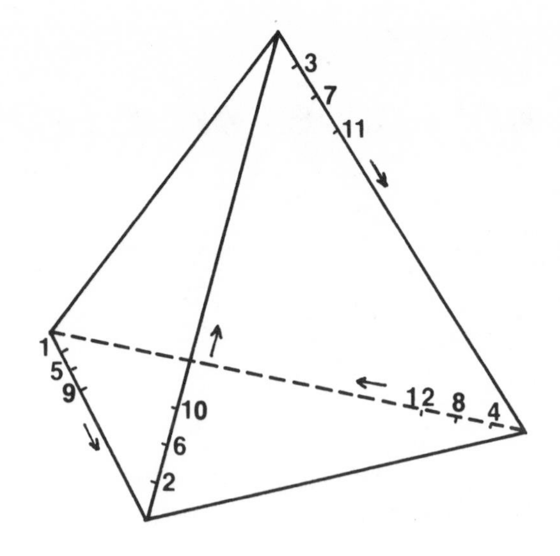

Fig. 3

Related Exercises

1. Each face of a tetrahedron is a (an) ______________________ .
2. Each face of a regular tetrahedron is a (an) ______________________ ______________________ .
3. In a tetrahedron, ______ edges meet at each vertex.

TEACHER NOTES

Activity 13

Grades 7–10

Geometric concepts
- Centimeter
- Face of tetrahedron
- Tetrahedron
- Interior of tetrahedron
- Parabola

Skills
- Measuring 1/2 cm and 12 cm
- Estimating 1 m of string
- Punching a straight row of holes through a straw
- Threading a needle
- Tying a knot

Answers to Related Exercises

1. Triangle
2. Equilateral triangle
3. 3

*ACTIVITY J

On Your Own

Two Curved Surfaces in a Tetrahedron

Based on Activity 11

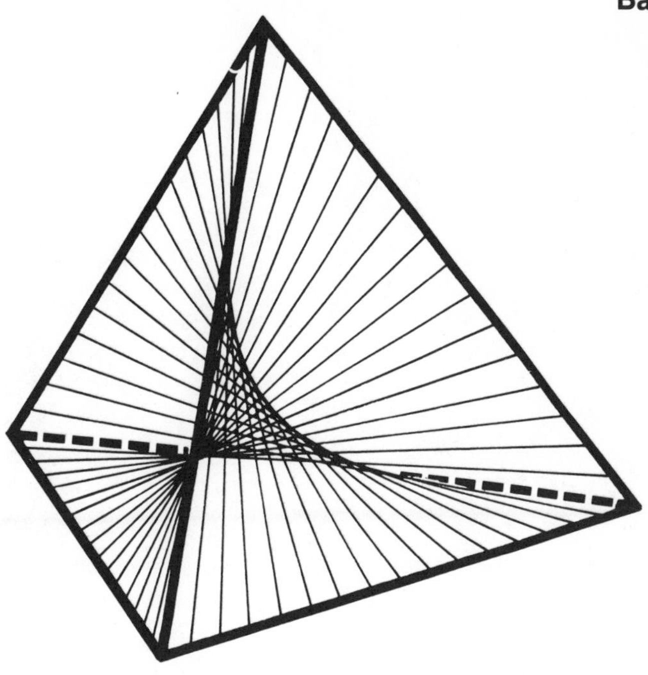

This design consists of two constructions of Activity 11—one on one pair of skew lines and a second on a second pair of skew lines. Weave each surface with a different color of string. If the strings seem to get in the way, try making one stitch of string 1 and then one stitch of string 2. Continue alternating the strings until the two curved surfaces are complete.

TEACHER NOTES

*Activity J

Grades 9 and above

See Activity 11.

TEACHER NOTES

Activity 14

Grades 8 and above

Since this construction takes quite a while, the teacher may wish to use several separate sessions. Any of the following may be considered a separate session:

1. Cut and punch the straws (step 1). (Punching 15 holes in 8 straws takes a bit of time.)
2. Build the octahedron (steps 2–5).
3. Weave the string (steps 6–9 or 10).

Geometric concepts
Octahedron
6 vertices of octahedron
12 edges of octahedron
Sets of parallel lines
Square in three-dimensional space
Intersecting planes (squares)
Perpendicular planes (squares)
Pyramid (half of octahedron)
Base of pyramid (a square)
Congruent pyramids
Interior and exterior of octahedron

Skills
Measuring 1/4-in. units
Measuring lengths of string
Approximating 3 in.
Punching straight rows of holes
Threading a needle

ACTIVITY 14

Two Squares in an Octahedron

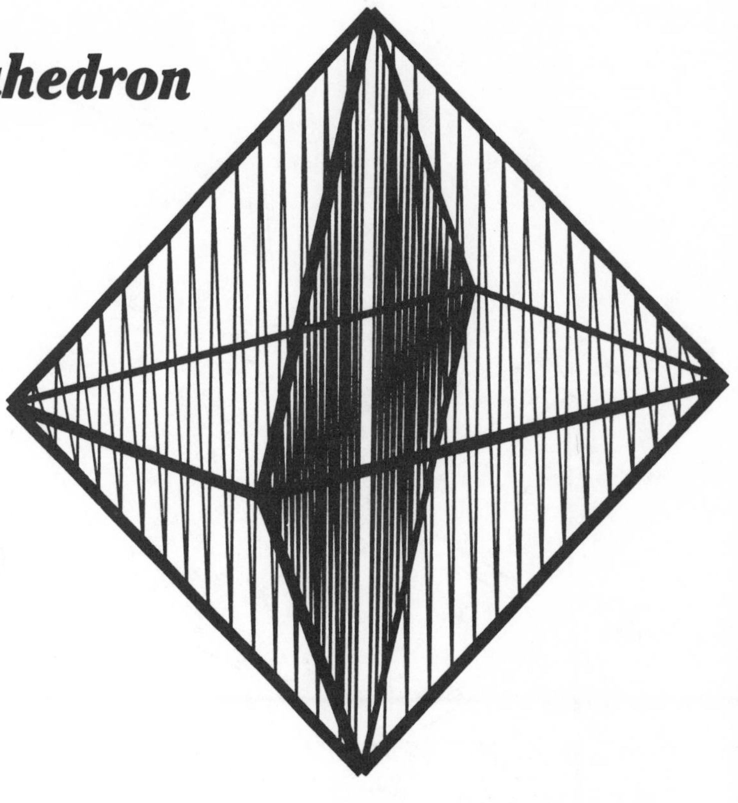

Materials

Ruler
Scissors
Glue
2 needles, one heavy and one rather thin
Plastic straws for cutting 12 lengths of 4 in.
3 colors of string:
  5 lengths of 3 ft. (color A) for inside the straws
  4 lengths of 4 1/4 ft. (color B) for first square
  4 lengths of 4 1/4 ft. (color C) for second square
Toothpick (optional for step 10)

Note: Wrap the strings on cardboard or rolled paper to prevent tangling.

Procedure

1. *Prepare the straws.* Cut the straws into twelve lengths of 4 inches. Use a needle or pin and a ruler to punch a straight row of fifteen holes 1/4 inch apart through a straw (fig. 1). Punch eight straws in this manner. Punching the holes is somewhat easier if done on a paper ruler.

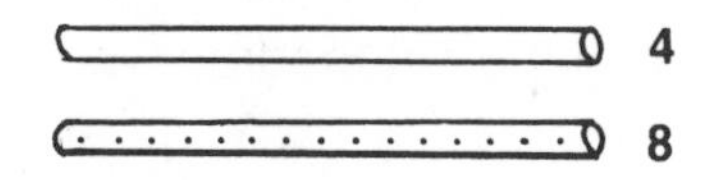

Fig. 1

2. *Build four triangles.* Thread the larger needle with a string of color A. Drop the threaded needle through a punched straw, then through a smooth straw and then through another punched straw. Pull and tie the string tightly (fig. 2). Cut the string about 3 inches from the tie. Now build three more triangles in exactly the same way.

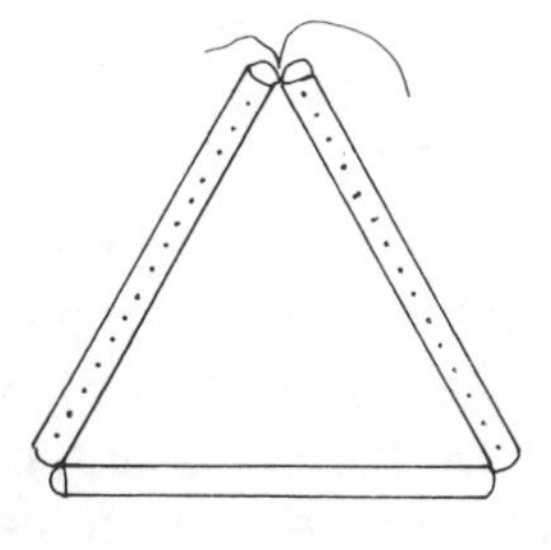

Fig. 2

3. *Tie pairs of triangles together.* Use the short strings at the top to tie two triangles together (fig. 3). Cut off two of the strings and keep two for later use. Tie the other two triangles together in the same way.

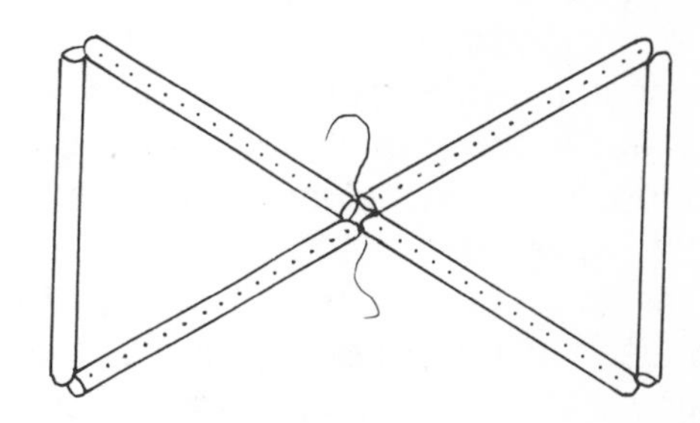

Fig. 3

4. *Build the octahedron.* Tie another string of color A to one of the tie strings on a pair of triangles. Thread the larger needle with this string. Drop the needle through the straws indicated by dark lines in figure 4. Follow the direction of the arrows. Pull the string tight, tie it, but do not cut it. Now drop the threaded needle through the "darkened" straws in figure 5, and slowly pull the string until it is secure. Tie it with a triple knot and cut off the unneeded string by this knot, but keep the tie strings at the other end. You now have an octahedron as shown in figure 6 with the right half sturdy and the left half a bit wobbly.

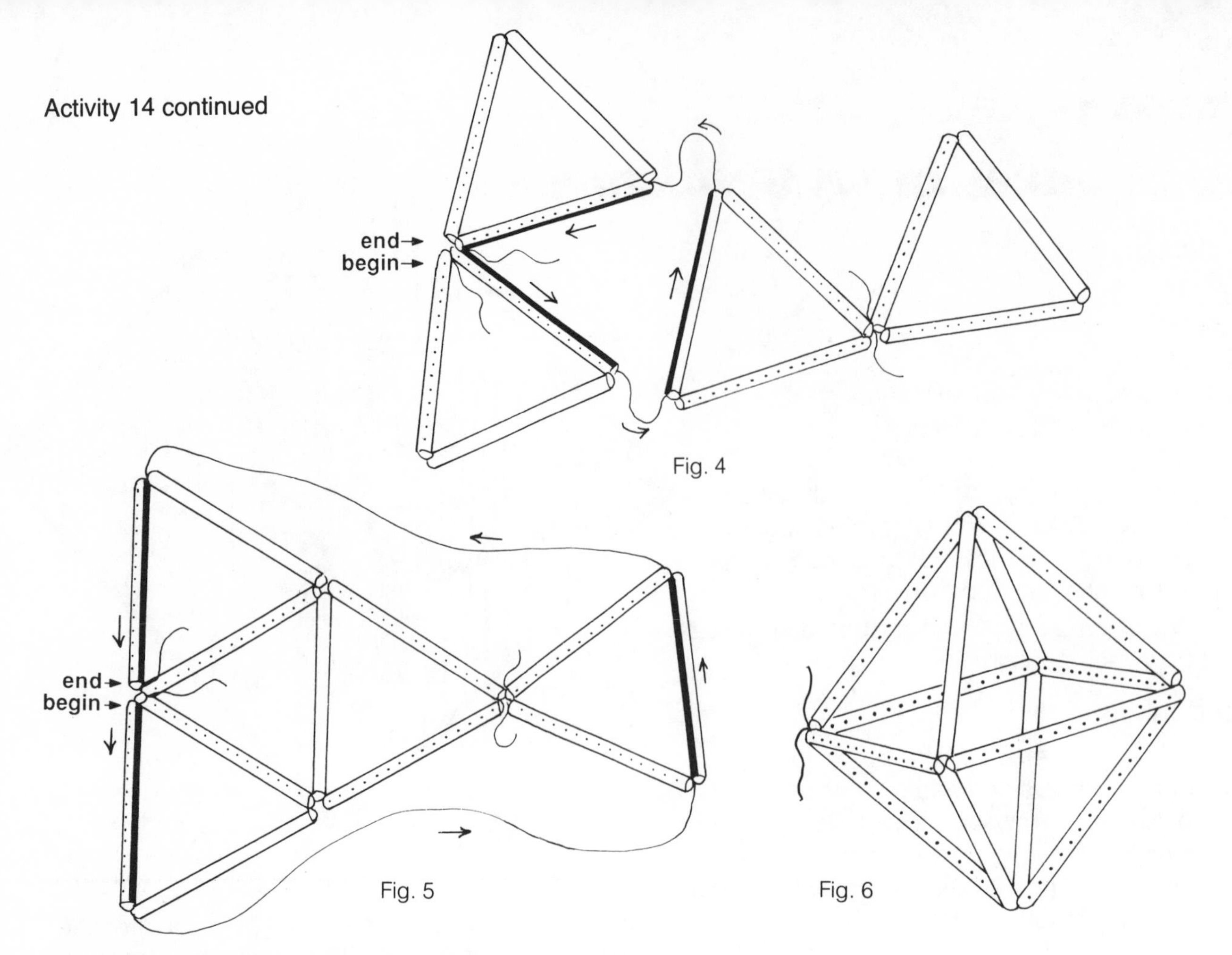

Fig. 4

Fig. 5

Fig. 6

5. *Stabilize the vertices.* In this step you will draw a total of three strings through each straw so the angles will not lose their shape during the weaving. Observe that two triangles are somewhat unattached, namely, those indicated by the darkened lines in figure 7. To join the unconnected straws, tie another string of color A to a tie string, thread it through the larger needle, drop it through the three straws of each triangle, then pull and tie the string. Now notice that three squares are formed by the twelve straws. Pass the threaded needle through the four straws of each of these squares, pull the string tightly, and tie it. Cut off any extra string. At each vertex you should now have all the connecting strings shown in figure 8. Gently position the punched straws so that smooth parts face the interior and exterior of the octahedron while holes face to either side.

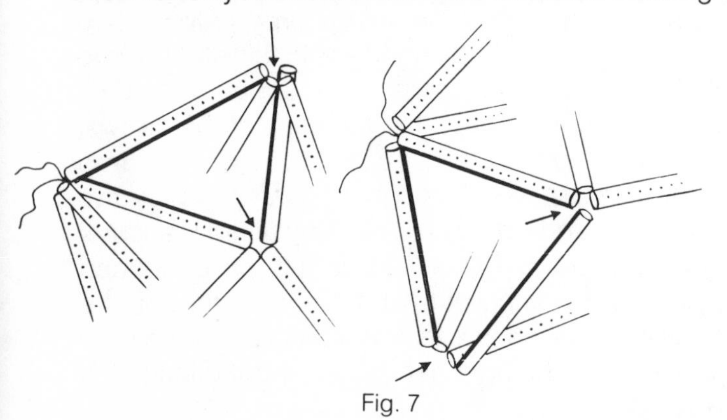

Fig. 7

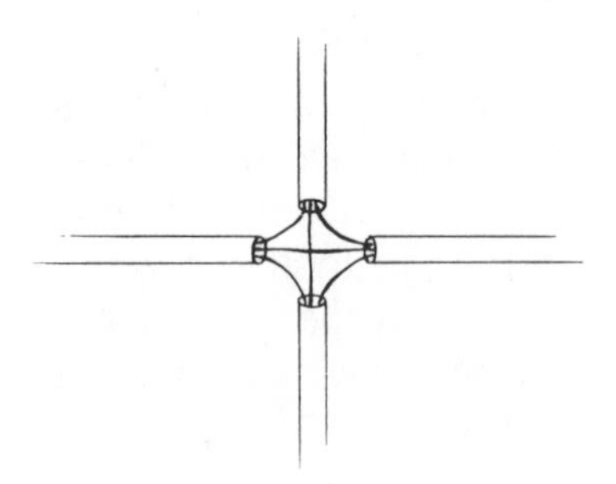

Fig. 8

6. *Weave the first half-square.* Thread the smaller needle with a length of the string you have chosen for the first square (color B). Knot the end of the string. To follow the numbering in figure 9, turn the octahedron so the tied vertices are at top and bottom. Connect the first few holes slowly and carefully because spacing is cramped near the vertex and you are learning the pattern. Push the threaded needle through hole 1. Next stretch the string down

to hole 2, then up to hole 3, down to 4, up to 5, down to 6, and so on. Pass the needle through the holes so that all the strings from top to bottom are on one side of the straws and those going back up are on the other side of the straws. When you are out of string, tie on another length of the same string so the knot is near a straw. Cut off short ends near the knot and proceed. The last two or three stitches must be done slowly because you will again be near a vertex. You need not use the last hole of each straw, since the straws would hide the tiny segment of string anyway. Do not tie the end of the string; cut off all but about 3 inches.

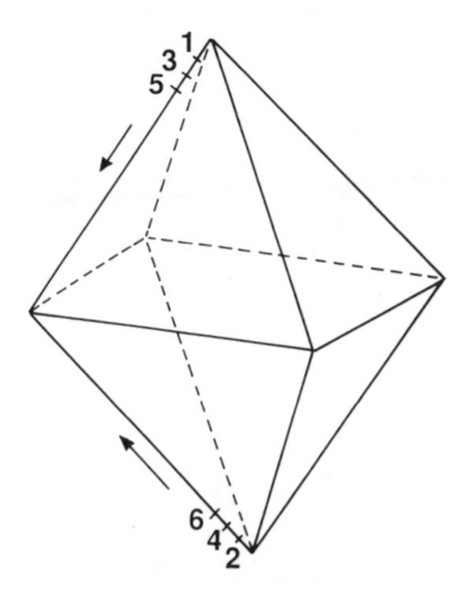

Fig. 9

7. *Weave the other half of the first square.* The procedure here is exactly the same as in step 6. Use the same color of string and the half-square shown in figure 10.

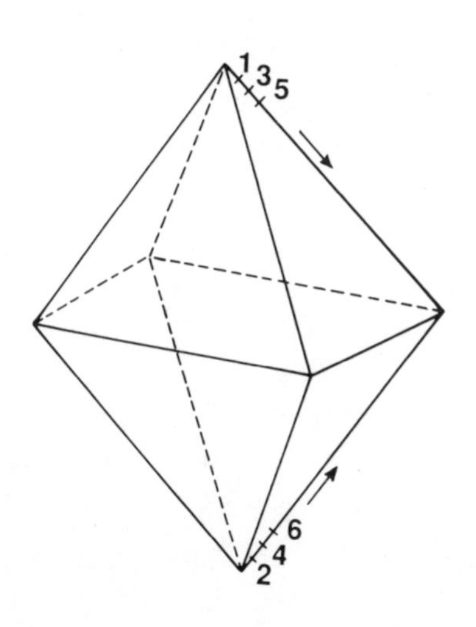

Fig. 10

8. *Weave the second square.* Now thread the smaller needle with the other color of string (color C). Follow the procedure described in steps 6 and 7, but use the half-squares shown in figures 11a and 11b. It does not matter which half-square you weave first. The lines of string in all four half-squares should extend up and down. While weaving the second square be careful not to pull the string too tightly or it will cause the strings in the first square to sag.

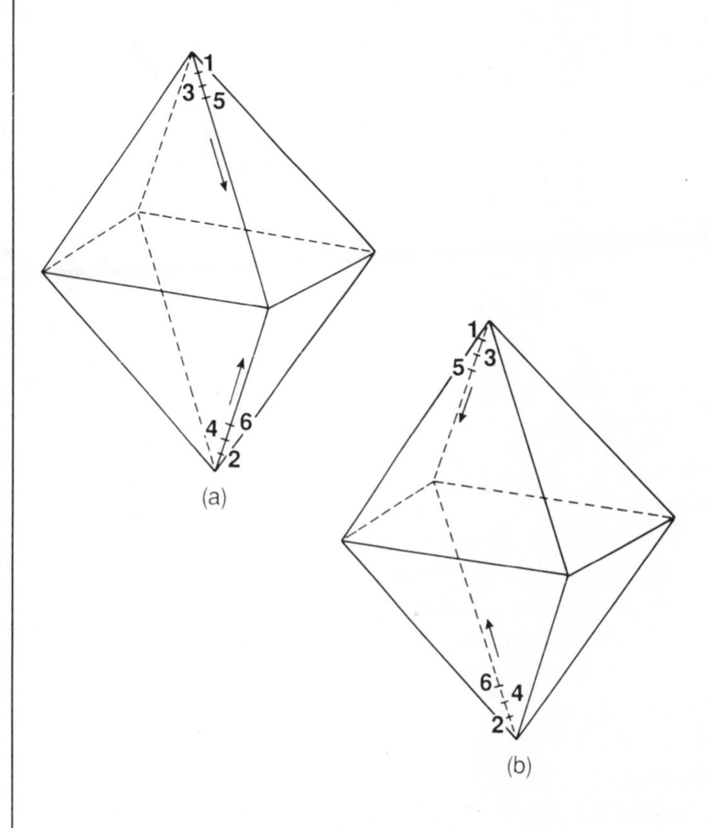

Fig. 11

9. *Straighten and tie the strings.* If any of the strings in your model are sagging, gently go over the sagging half-square beginning with the sagging string and working toward the outside. When all strings form straight lines, knot each string close to a straw and cut off the extra string. Glue each knot against a straw to make the knots less noticeable.

10. (Optional) *Glue the vertices.* Put glue on the inside of each of the six vertices. With a toothpick spread the glue between the straws so that neighboring straws are glued together. Push the straws a little so each vertex is neatly arranged.

Teacher Notes for Activity 14 on page 48

ACTIVITY K **On Your Own**

Two Footballs in an Octahedron

Based on Activities 5 and 14 or 16

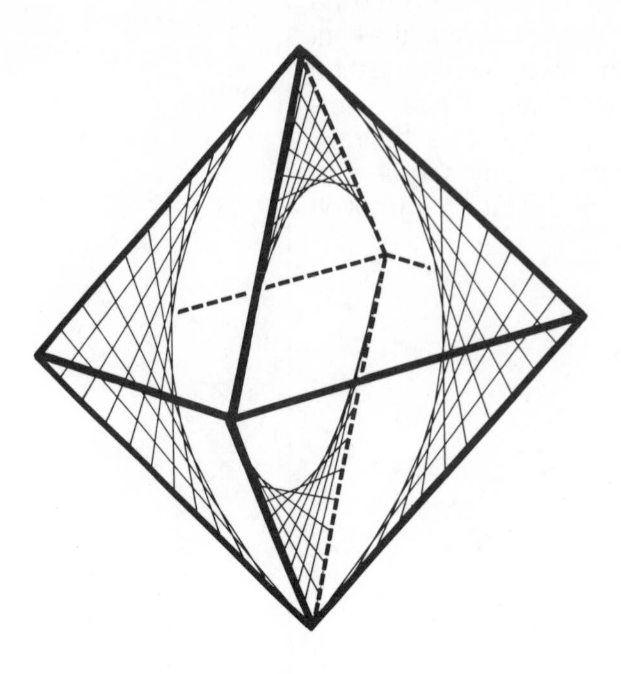

Study the design and then punch straws or notch wood dowels. Next use Activity 14 or 16 as a guide to construct an octahedron. Find a square formed by four edges of the octahedron. Then weave a football in each square using Activity 5 as a guide.

TEACHER NOTES

Activity K

Grade 8 and above

See Activity 14 or 16 for octahedron and Activity 5 for string design.

ACTIVITY 15

Eight Curves on an Octahedron

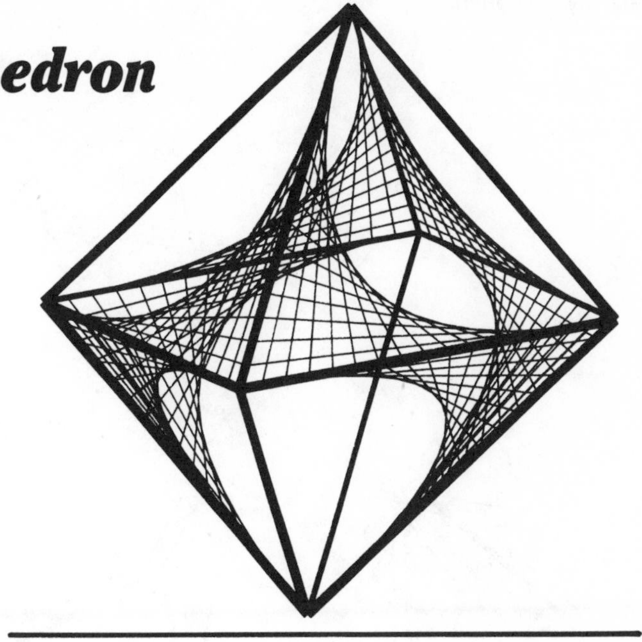

Materials

Ruler
Scissors
Toothpick
Glue
String, 45 ft.
Plastic straws for cutting 12 lengths of 4 in.
2 needles, one heavy and one rather thin

Procedure

1. *Prepare the straws.* Cut twelve straws 4 inches long. Use a needle or pin and a ruler to punch a straight row of holes 1/4 inch apart through a straw. Punch eight straws in this manner.

2. *Build four triangles.* Thread the larger needle with about 3 feet of string. Drop the threaded needle through two punched straws and then through a smooth straw. Next pull and tie the string to form a triangle with no slack in the string (fig. 1). Cut the string about 3 inches from the end. Now build three more triangles in exactly the same way.

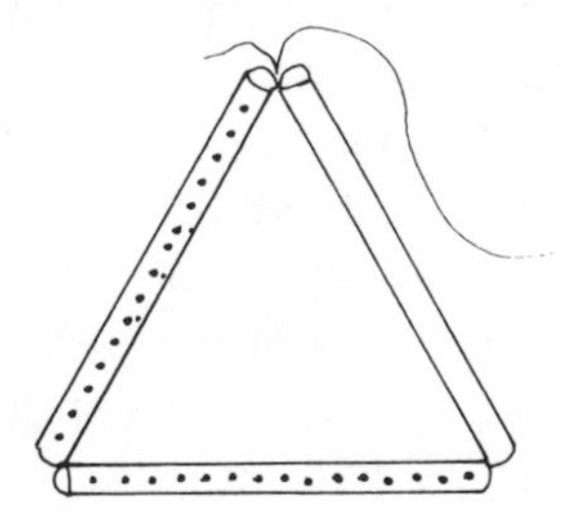

Fig. 1

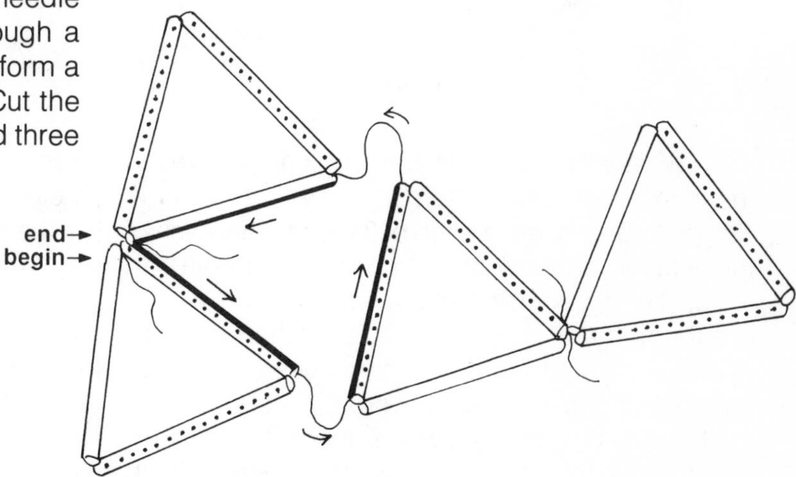

Fig. 3

3. *Build the octahedron.* Follow steps 3, 4, and 5 of Activity 14, except that the positions of the punched straws will be different. Use the arrangements shown here in figures 2–6.

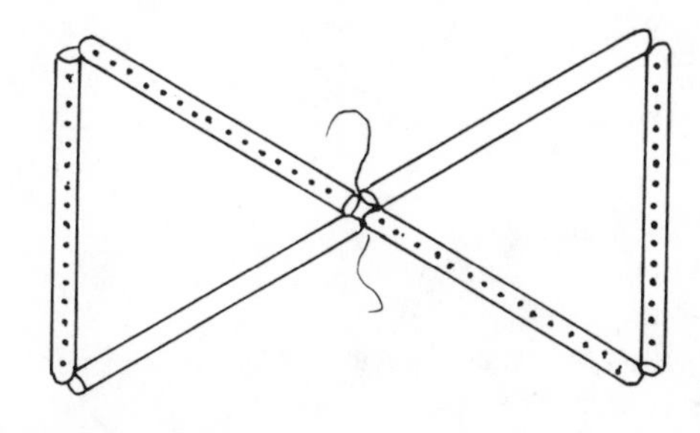

Fig. 2

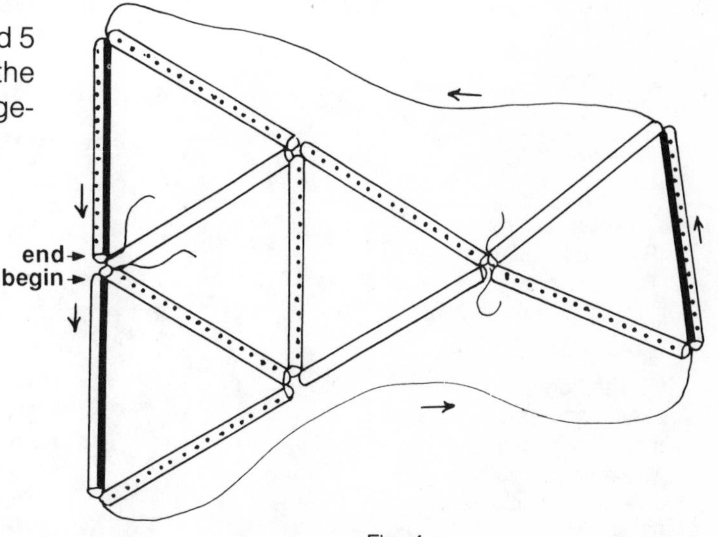

Fig. 4

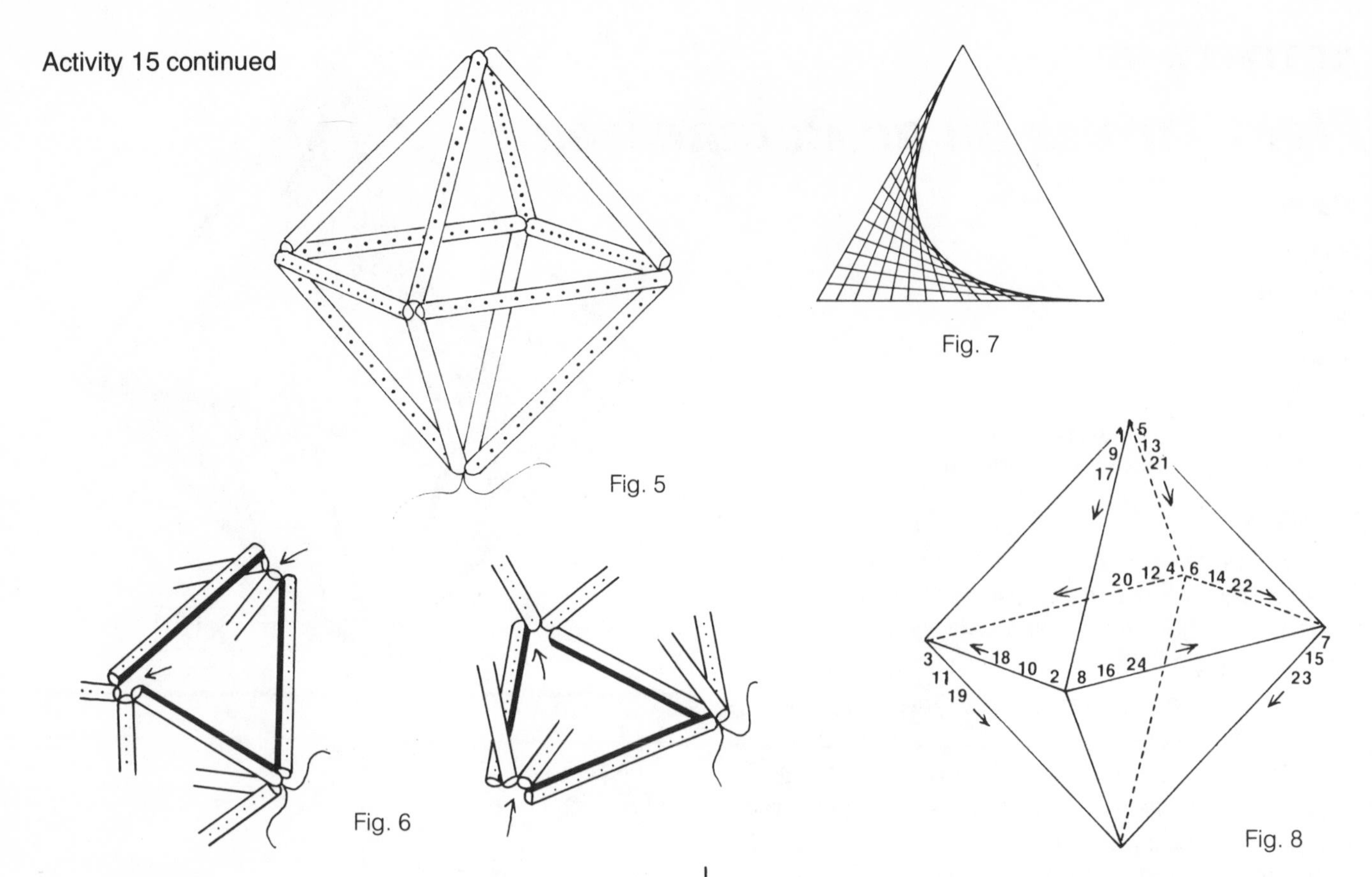

Fig. 5

Fig. 6

Fig. 7

Fig. 8

4. *Weave the curves.* Before beginning to weave the string, study the pattern in figure 7. This is the design to be woven in each of the eight triangles of the octahedron. Now thread the smaller needle with 4 feet of string. Knot the end of the string so it will not slip through a hole. Follow the numbering in figure 8. Make the first eight to ten stitches slowly and carefully. Push the needle through hole 1. Next stretch the string down to hole 2, then over to hole 3, over to 4, up to 5, down to 6, over to 7, over to 8, up to 9, and so on. (The drawing shows only three holes on each punched straw to avoid a clutter of numbers.) Each time the needle passes through a hole, pull the string securely but not too tightly or it will arch the straws.

When all the string is used, attach another 4 feet of string in such a way that the knot is on a straw or as near one as possible. Continue weaving the string in this manner until the needle has passed through the last hole. Then tie a knot and cut off the unneeded string. Glue each knot against a straw to make the knots less noticeable. Your straw octahedron containing eight parabolas on its faces is now complete.

TEACHER NOTES

Activity 15

Grades 8 and above

Three sessions are recommended:

1. Prepare the straws (step 1).
2. Build the octahedron (steps 2 and 3).
3. Weave the design (step 4)

Geometric concepts

Octahedron
Interior and exterior of octahedron
8 faces of octahedron
12 edges of octahedron
6 vertices of octahedron
Parabola

Skills

Measuring 1/4-in. units
Measuring lengths of string
Approximating 3 in.
Punching straight rows of holes
Threading a needle

ACTIVITY 16

Four Curved Surfaces in an Octahedron

Materials

Ruler
Glue
Pencil
Tape
String, 34 ft.
12 wood dowels, 1/8 in. in diameter and 5 in. long
Small knife or file with filing edge
Small saw or pliers with center cutting section
For optional step 5: toothpick
For optional step 6: paint

Note 1: Wrap the string on a small card (about 2 by 3 inches) because it will have to be flat near the end of the weaving.

Note 2: This activity may also be done on 5-inch straws. In that case begin with four triangles like the one pictured and follow Activity 14 to build the octahedron. Then proceed to step 7 for the weaving instructions.

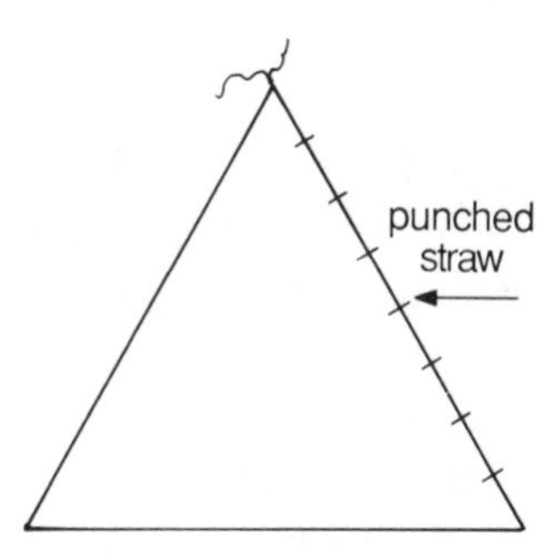

Procedure

1. *Cut and mark the sticks.* Cut twelve wood dowels, each 5 inches long. It is helpful but not necessary to slice both ends of each stick on the diagonal with a knife or file (fig. 1a). Note that the long part is at the top on both ends. With a pencil, mark every 1/4 inch on four sticks. Use the longest side if you slanted the ends of the stick (fig. 1b).

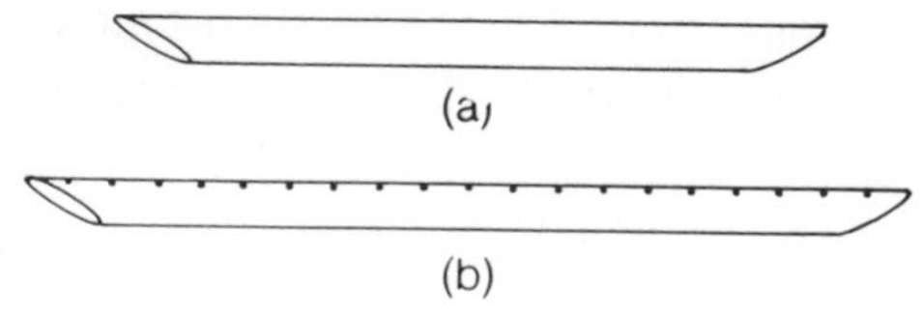

(a)

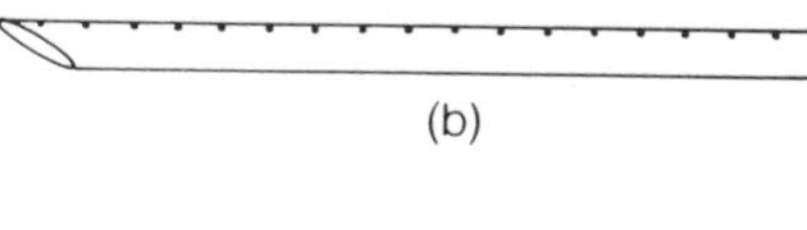

(b)

Fig. 1

2. *Make the notches.* With a knife or the thin filing edge of a file, make a notch at each of the marks (fig. 2). Try to cut or file the notches in a straight line. They must be deep enough to hold one string.

Fig. 2

3. *Build two pyramids.* Glue four unnotched sticks together to form a square (fig. 3). Next apply glue on both ends of two notched and two smooth sticks. When this glue is thick and sticky, arrange the four sticks on the square (with the notched sticks nonadjacent) to form a pyramid (fig. 4). You may want someone to hold the first two sticks in place while you position the other two. Arrange all five vertices neatly, and position the sticks so that notches (and the longest part of the smooth sticks) face the exterior. Next apply glue on one end of each of the last four sticks. Attach 1-inch pieces of tape to the dry end of each stick. On a piece of scrap paper, draw a square with each side 5 inches. When the glue on the last four sticks is stiff,

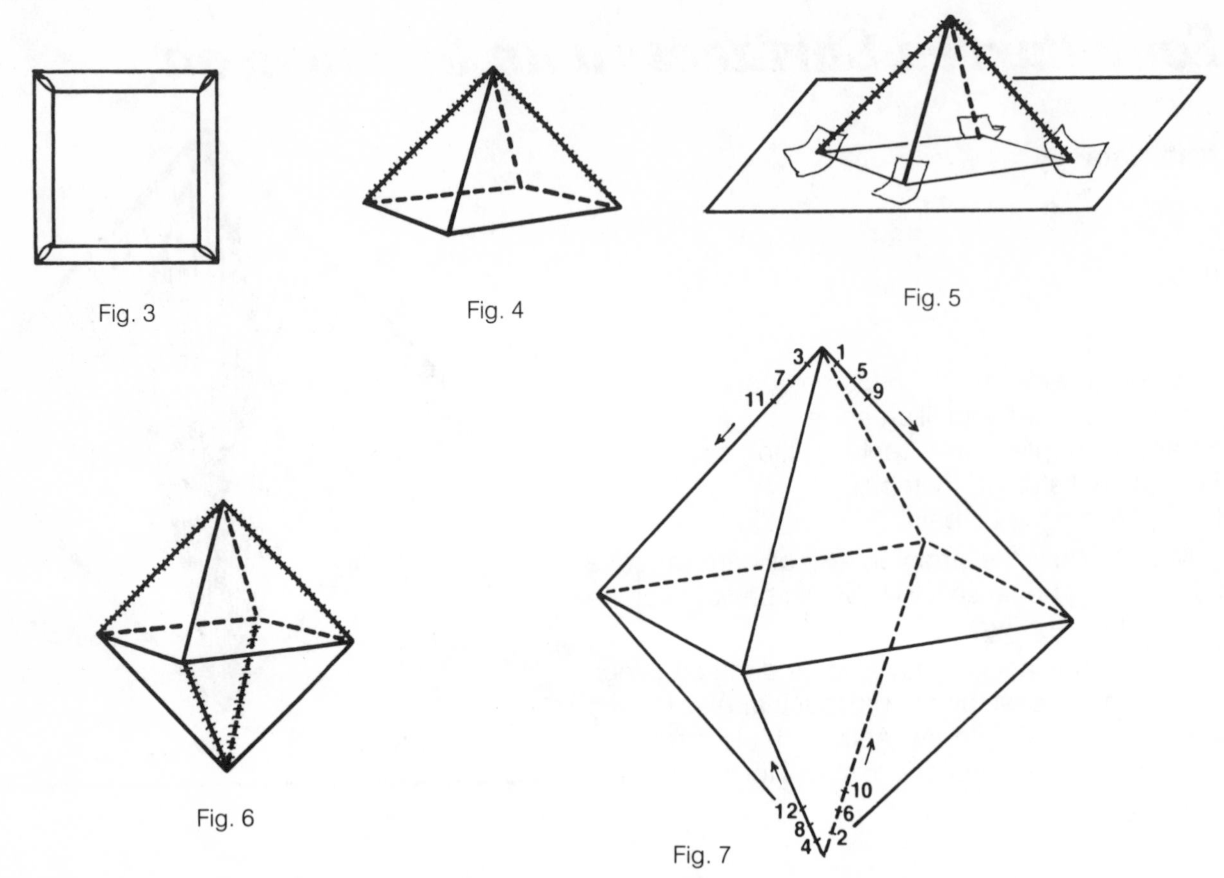

Fig. 3

Fig. 4

Fig. 5

Fig. 6

Fig. 7

build a pyramid on the paper square (fig. 5) by taping the ends of sticks to the corners of the square and having the glued ends form the vertex. (Again, notched sticks are to be nonadjacent; notches and long sides face the exterior.) Allow both pyramids to become set before proceeding to step 4.

4. *Build the octahedron.* Remove the tape from the four sticks. Glue the four-stick pyramid onto the eight-stick pyramid to form the octahedron shown in figure 6. Note that the notched sticks of one pyramid meet the smooth sticks of the other pyramid. Gently push the four sticks so they fit onto the corners of the square. Allow the glue to dry.

5. (Optional) *Touch up the vertices.* At each of the six vertices use a toothpick to fill the holes and cracks with glue. This includes both the interior and the exterior of each vertex. The glue will give the model a finished look and will also strengthen it.

6. (Optional) *Paint the octahedron.* If you wish to paint your wooden figure, this is the time to do it. Paint is not necessary, but if you use it, choose paint that will not clog up the notches and a color that will go well with the strings.

7. *Weave the surfaces.* When the glue (and paint) are thoroughly dry, tie or glue the end of your string to notch 1. Follow the numbering in figure 7. Stretch the string down to notch 2, then back up to 3, down to 4, and back to notch 5 on the first stick, then on to 6, 7, 8, 9, and so on. Keep the string taut at all times. Glue the string in every fourth or fifth notch so there will be no big problem if the string or octahedron slips while you are weaving. Continue weaving until you have used all the notches. Finish by gluing the string in the last notch; then extend the string along the underside of an adjacent stick, gluing it at the stick's midpoint. This completes the pattern of almost-parallel strings. Cut off the extra string. You now have four curved surfaces in your octahedron, and each surface connects two skew lines.

Related Exercises

1. **Octa** means 8 and **hedron** means face. Hence, an octahedron has ______ triangular ____________________ .
2. An octahedron also has ______ edges and ______ vertices.
3. ________ edges meet at each vertex of an octahedron.
4. Each edge of an octahedron intersects ______ edge(s), is parallel to ______ edge(s), and is skew to the remaining ______ edge(s).
5. A regular octahedron has ______ lines of symmetry. ______ of these connect opposite vertices; ______ connect centers of opposite faces; and ______ connect midpoints of opposite edges.
6. Create your own string design on an octahedron.
7. A curved surface connects two ____________________ lines; a parabolic curve connects two ____________________ lines.

TEACHER NOTES

Activity 16

Grade 9 and above

If possible, cut the sticks the desired length ahead of time either in the school's industrial arts department or in a home workshop.

Three sessions are recommended:

1. Prepare the sticks (steps 1 and 2).
2. Build the octahedron (steps 3–6).
3. Weave the design (step 7).

Geometric concepts
Octahedron
Pyramid
Interior and exterior
Adjacent and nonadjacent edges
Skew lines
6 vertices of octahedron
5 vertices of pyramid
Curved surface

Skills
Measuring 1/4-in. units
Drawing a square
Filing or cutting straight rows of notches
Gluing sticks together

Answers to Related Exercises

1. 8; faces
2. 12; 6
3. 4
4. 6; 1; 4
5. 13; 3; 4; 6
7. Skew; intersecting

ACTIVITY L

On Your Own

Eight Raindrops on an Octahedron

Based on Activities 1 and 14 or 16

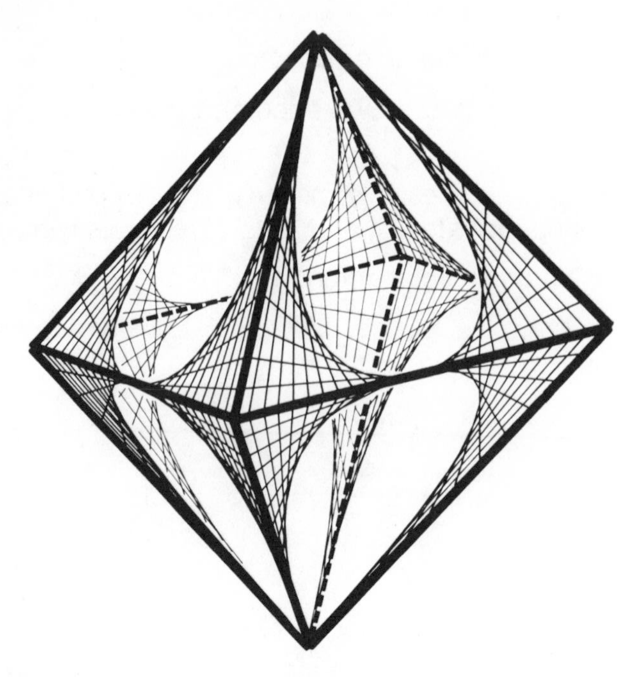

It is possible to weave a raindrop on each of the eight faces of an octahedron using each edge only once. Can you discover a systematic way to do this? The raindrop pattern is given in Activity 1. Use either straws or wood dowels.

TEACHER NOTES

Activity L

Grade 9 and above

See Activity 14 or 16 for octahedron and Activity 1 for string design.

**ACTIVITY 17

Ten Curves on an Icosahedron

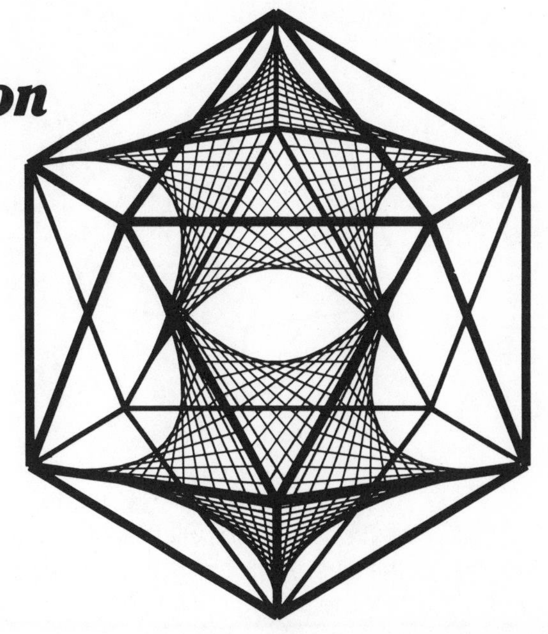

Materials

Centimeter ruler
Glue
Pencil
String, two lengths, each 4.6 m long, wound on spools or cardboard and unwound as it is used
30 wood dowels, 1/8 in. in diameter and 8 cm long
Small knife or file with filing edge
Small saw or pliers with center cutting section
Optional materials for step 5: toothpick and paint

Note: This activity may also be done with straws, using approximately the same process as that suggested for dowels. Keep the string taut, tie it frequently, and make five straws meet at each vertex. Tighten up the icosahedron a bit by dropping the threaded needle through straws to connect them with their neighbors. Glue the vertices before beginning to weave the design.

Procedure

1. *Prepare the sticks.* Slant the ends of the dowels if you wish, as in Activity 16. Mark every 1/2 cm on ten sticks. If you do not have a centimeter ruler, use the paper ruler below. Make a notch at each mark (fig. 1).

Fig. 1

2. *Build two pyramids.* Glue together unnotched sticks to make two pentagons (fig. 2). Apply glue to both ends of all notched sticks. Let the glue on the sticks get stiff and then build the pyramids by using a pentagon for the base and your construction of notched sticks for the slanted sides (fig. 3). (Notches should face the exterior of the pyramid.) Push all the vertices into a good shape, and allow the pyramids to dry.

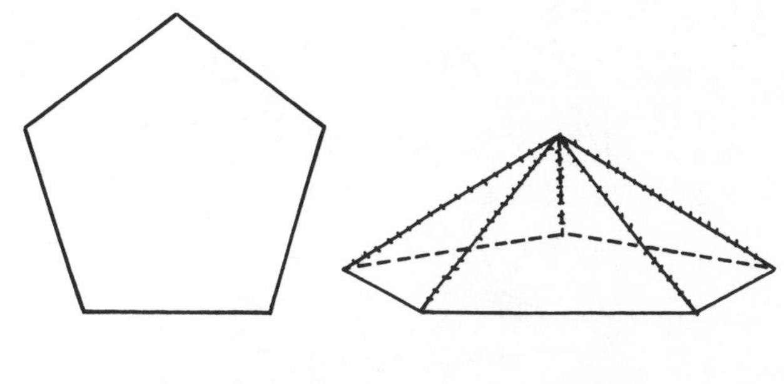

Fig. 2 Fig. 3

3. *Stellate one pyramid.* Turn the pyramid upside down, bracing it in a tin can. Apply glue to both ends of the last ten sticks. (Doing a few at a time is probably best.) When the glue is rather stiff, form five triangles along the base of one of the pyramids as shown in figure 4. This is a difficult step and is made a little easier by bracing the triangles with little boxes or tin cans. The triangles should lean slightly outward. When possible, push the sticks together to form neat vertices. Allow this figure to set but not to get thoroughly dry.

1	2	3	4	5	6	7	8	9	10	11	12	13	14	15	16	17	18

Centimeter ruler

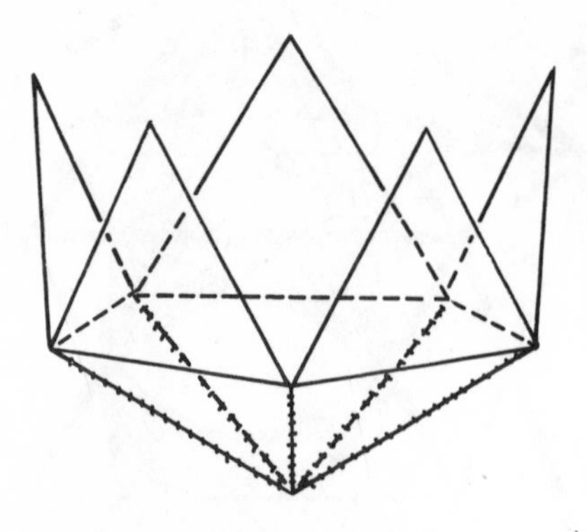

Fig. 4

4. *Build the icosahedron.* Glue the tips of the five triangles to the five vertices of the other pyramid. Carefully push the triangles into position one by one. They will form a rigid figure like the one shown in figure 5. This figure is called an icosahedron, a name meaning twenty faces. Notice that the icosahedron does have twenty triangular faces.

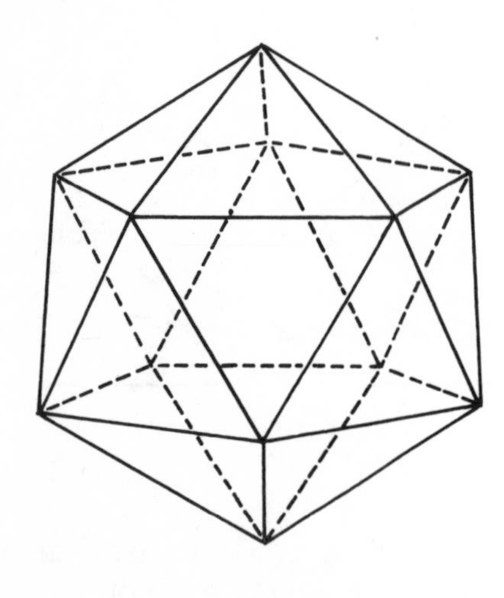

Fig. 5

5. (Optional) *Touch up the icosahedron.* You are encouraged to fill in the vertices with glue. Paint the wooden figure if you wish. (See steps 5 and 6 of Activity 16.)

6. *Weave five curves.* Glue (and paint) must be thoroughly dry before the weaving begins. Find a vertex surrounded by five notched edges. Begin at this vertex on notch 1 and follow the numbering in figure 6. Tie or glue the end of the string to notch 1. Now stretch the string through notches 2, 3, 4, 5, 6, 7, and so on, in that order. The pattern results from weaving from both ends of the edges toward their midpoints. Glue the string into notches as needed. Continue weaving until you have glued string in the last notch (a midpoint). Then stretch the string to the midpoint of the next edge to fill in the empty space. Glue the string in the middle notch and cut off unneeded string.

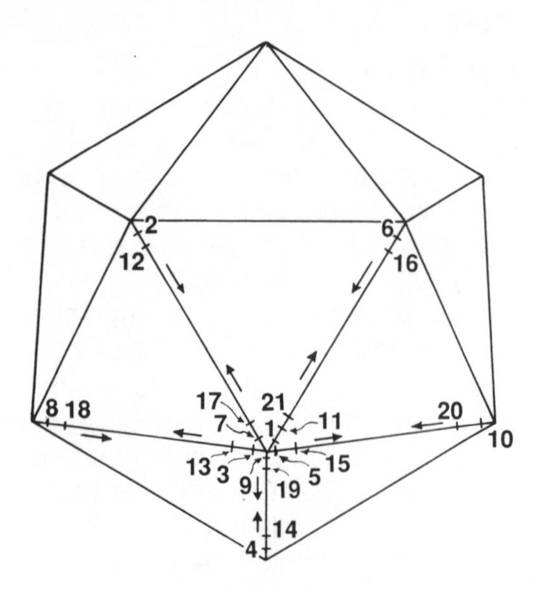

Fig. 6

7. *Weave five more curves.* Now repeat step 6 on the other set of five notched edges. When you have completed this step, you have woven ten parabolas on an icosahedron.

Optional: If you would like to have a more intricate string design on your icosahedron, weave parabolas on the middle section of the icosahedron as shown in figure 7.

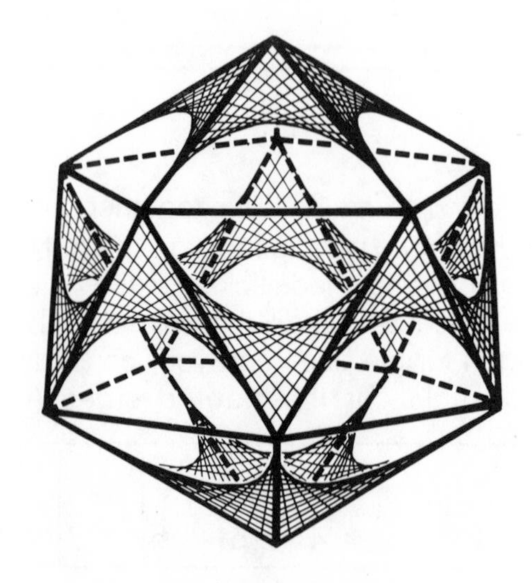

Fig. 7

Related Exercises

1. An icosahedron has ______ faces, ______ edges, and ______ vertices.
2. ________ edges meet at each vertex of an icosahedron.
3. On an icosahedron, weave a design similar to the string pattern in Activities 13 and 15. That is, weave a parabola on each face so that these twenty parabolas form one continuous band around the icosahedron. (This *is* possible.)
4. On an octahedron, weave a design similar to the string pattern in Activity 17.

*5. A regular icosahedron has ______ lines of symmetry. ______ of these connect midpoints of opposite edges; ______ connect centers of opposite faces; and ______ connect opposite vertices.

*6. Each edge of an icosahedron meets ______ edges and is parallel to ______ edge(s). Therefore it is skew to the remaining ______ edges.

TEACHER NOTES

**Activity 17

Grade 10 and above

Three or four sessions are recommended:

1. Prepare the sticks (step 1)
2. Build the icosahedron (steps 2–5). (This might take two sessions.)
3. Weave the curves (steps 6 and 7).

Geometric concepts
Icosahedron
Pentagonal pyramid
Parabola
Base of pyramid
Exterior of pyramid
20 faces of icosahedron
30 edges of icosahedron

Skills
Measuring 1/2 cm units
Filing or cutting straight rows of notches
Gluing sticks together in difficult positions

Answers to Related Exercises

1. 20; 30; 12
2. 5
3.

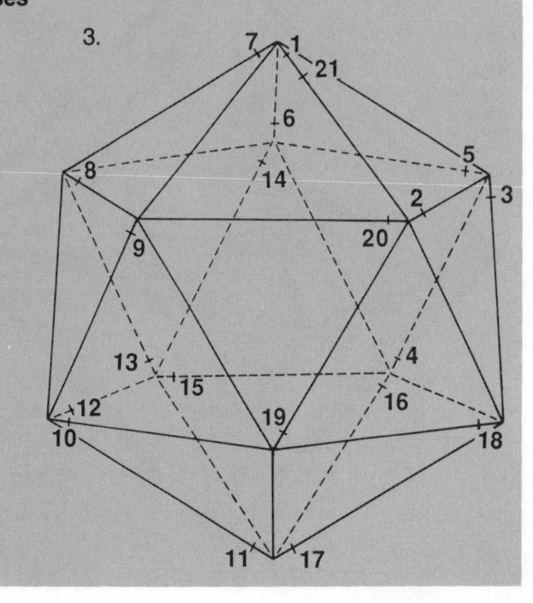

5. 31; 15; 10; 6
6. 8; 1; 20

*ACTIVITY M

On Your Own

Unions of Designs

Each of these exceptionally beautiful models consists of two previously introduced designs on one polyhedron. They can be done on wood dowels or straws. Use two colors of string and weave the interior design first. The small figures show which edges to notch or punch but do *not* indicate the correct number of notches.

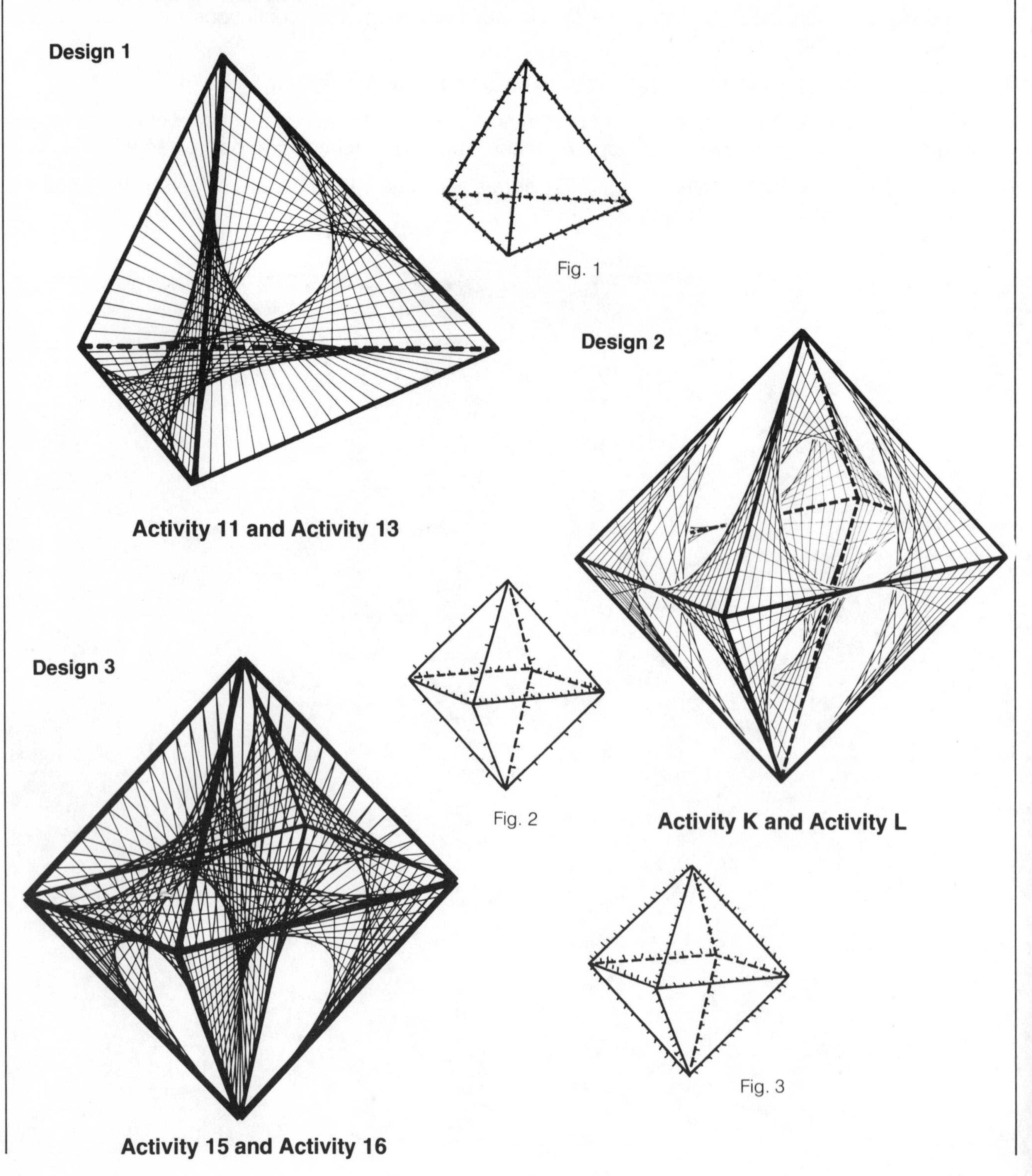

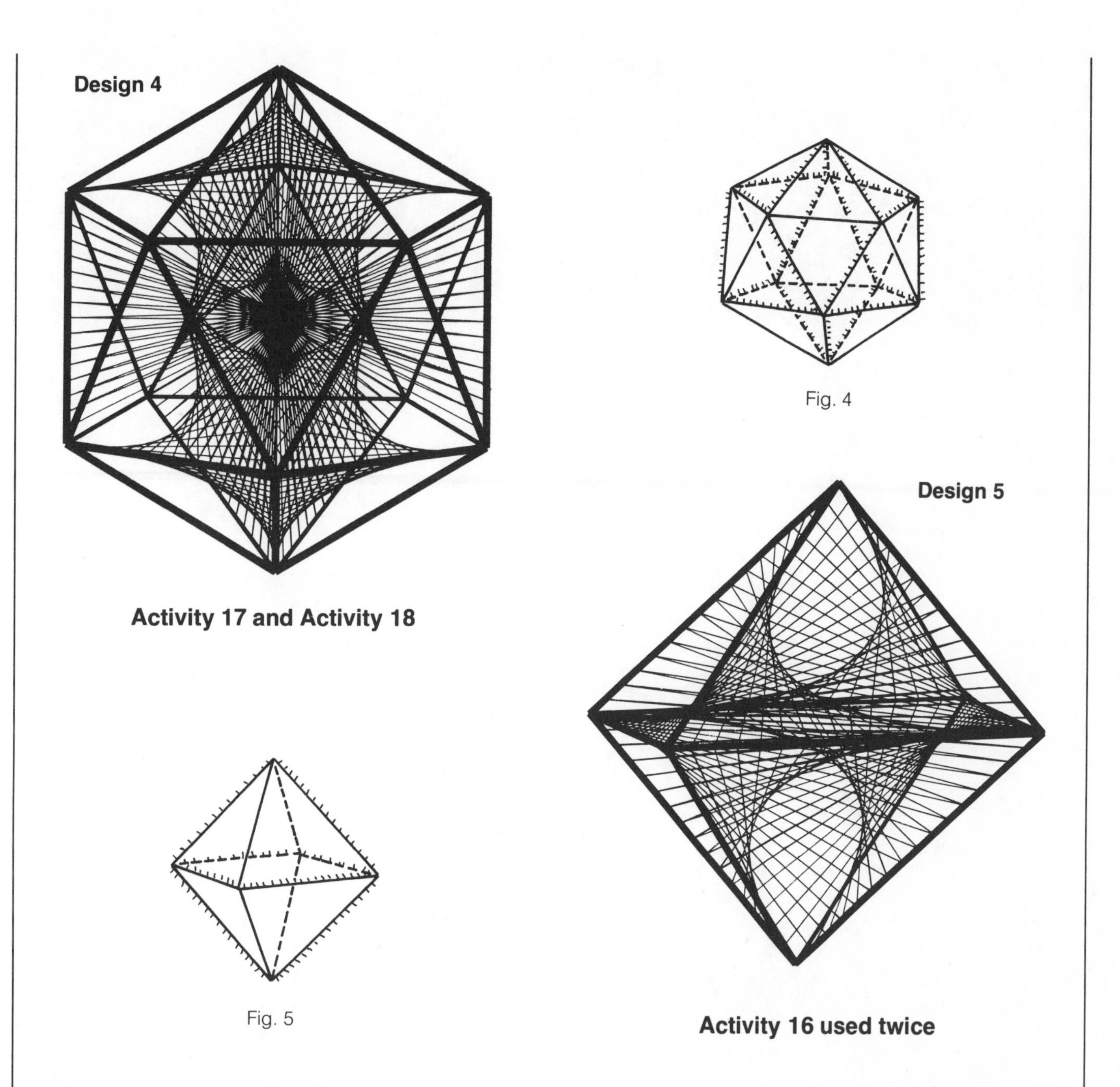

Hint for Design 5: Weave through four notches with one string, then four notches with the other string. Continue to alternate strings.

TEACHER NOTES

***Activity M**

Grade 7 and above

Design 1: 7–10
Design 2: 9 +
Design 3: 9 +
Design 4: 10 +
Design 5: 10 +

See Activities 11 and 13, K and L, 15 and 16, 17 and 18, or 16. These designs are magnificent. Try them; you and your students will like them.

**ACTIVITY 18

Concurrent Lines on Parallel Edges of an Icosahedron

Materials

Ruler
Glue
Pencil
Small knife or file with filing edge
Small saw or pliers with center cutting section
30 wood dowels, 1/8 in. in diameter and 3 in. long (or plastic straws)
5 lengths of string, each 11 ft. long (one or five colors)
Copy of Activity 17
Optional materials for step 2: toothpick and paint

Note: Wind each string on a piece of cardboard about 1 inch wide and 6 inches long.

Procedure

1. *Prepare the sticks.* Cut the dowels or straws. Mark 10 sticks every 1/4 inch and notch at each mark.

2. *Build the icosahedron.* Follow the directions in steps 2–5 of Activity 17 but change the placement of the notched sticks as follows: Build both pyramids using smooth sticks; then use the ten notched sticks to stellate one pyramid.

3. *Weave on two parallel lines.* Choose any of the notched edges and find the one edge that is parallel to it. This edge is on the opposite side of the icosahedron, has the same slant as the first edge, and is also notched. These two parallel edges are called k and k' in figures 1 and 3. Now follow the

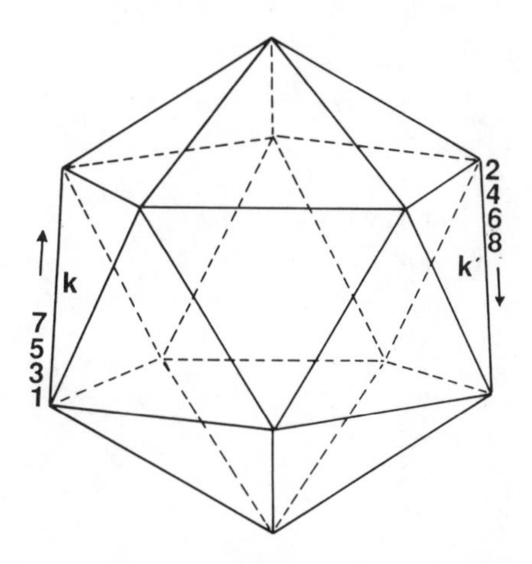

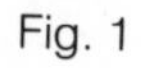
Fig. 1

numbering in figure 1. Glue your first string in notch 1 so that about 1 inch of string extends downward. This short end will be used as a marker for weaving step 3. Throughout steps 3–5 the string is to enter each notch from below and leave from above as shown in figure 2. This is done so that every string passes as near as possible to the center of the icosahedron. Pull the string over edge k, through the interior of the icosahedron, under edge k', and through notch 2. Then draw the string back to notch 3 in a similar manner. When passing the string through the interior, always stretch the string on the same side of the center as the 1-inch string. Hence strings from k to k' pass above k, then under the center, then under k'; and strings from k' to k pass above k', then under the center, then under k. To keep the string from slipping, glue it in notches as necessary. Continue weaving through notches 4, 5, 6, 7, and so on, until you have glued the string in the last notch of the two parallel edges. Cut off unneeded string and the 1-inch end.

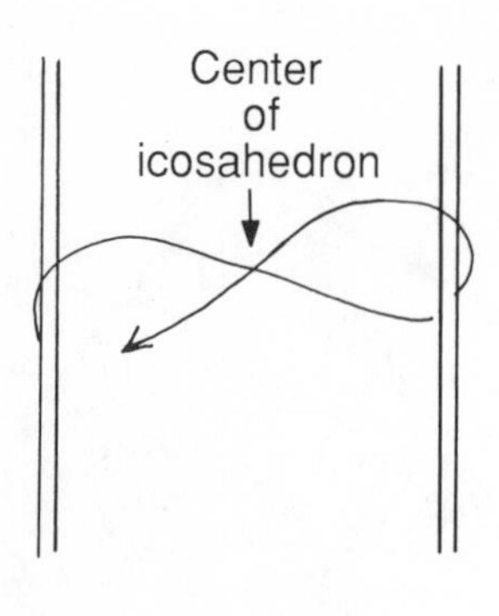

Fig. 2

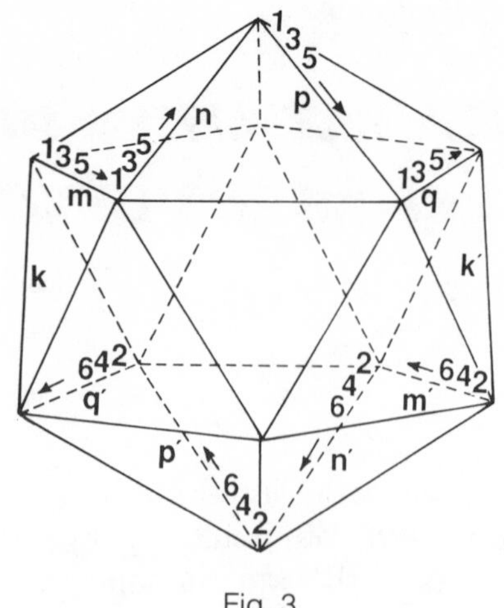

Fig. 3

4. *Weave on a second pair of parallel lines.* Now move to an adjacent notched edge and the edge to which it is parallel (m and m' in figure 3). Follow the winding instructions of step 3 for lines m and m', but the 1-inch string is not needed. Instead, all the strings on m and m' should pass through the interior of the icosahedron below the strings on k and k'; that is, keep all the m-m' strings together on one side of the previous design. This will cause the overlap of strings to gradually form a pattern at the center. So for steps 4 and 5, stretch the string from one edge to the next by passing the string over an edge, under the central design, and then under the opposite edge.

5. *Weave on the other six notched edges.* Now weave the same design on the parallel edges n and n', then p and p', and lastly q and q'. Each time, all strings on a pair of parallel edges are to pass through the center below the strings from previous steps. The last strings will be somewhat crowded because they will be near the first strings. A little mistake will not ruin the design, but try to keep all strings from p and p' together as they pass the center.

You have now woven 110 lines concurrent at the center of an icosahedron. That is, all of these lines meet at the center. Furthermore, these concurrent lines are located on five pairs of parallel edges of the icosahedron. That is impressive!

TEACHER NOTES

**Activity 18

Grade 10 and above

Three or four sessions are recommended:

1. Prepare the sticks (step 1).
2. Build the icosahedron (step 2). (This might take two sessions.)
3. Weave the curves (steps 3–5).

Geometric concepts

Icosahedron
Pentagonal pyramid
Base of pyramid
Exterior of pyramid
20 faces of icosahedron
30 edges of icosahedron
Adjacent edges
Center of icosahedron
Concurrent lines (the strings)
Parallel lines in space (the parallel edges)
Naming line segments with lowercase letters
Interior of icosahedron

Skills

Measuring 1/4-in. units
Filing or cutting straight rows of notches
Gluing sticks together in difficult positions

ACTIVITY N **On Your Own**

Concurrent Lines on Parallel Edges of an Octahedron

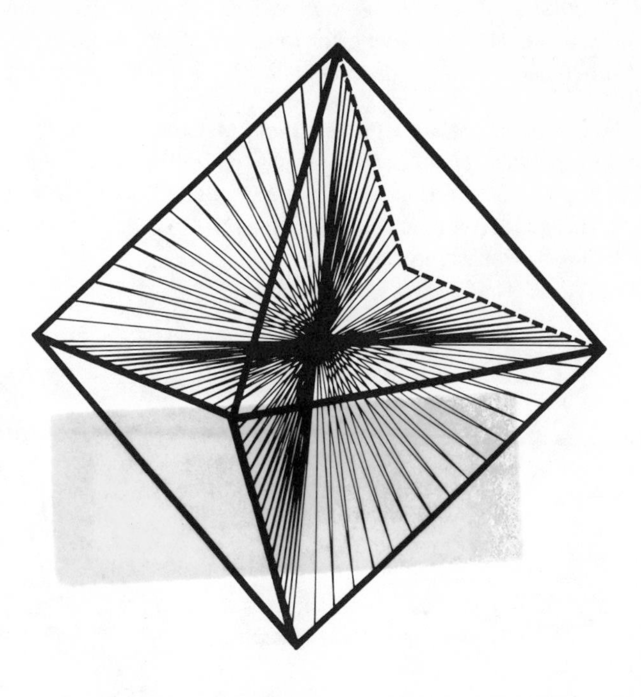

Based on Activities 14 or 16 and 18

Prepare six wood dowels or straws and then position them to form three pairs of parallel edges on an octahedron. Use the general idea of the string design described in steps 3–5 of Activity 18 to weave lines intersecting at the center of the octahedron.

TEACHER NOTES

Activity N

Grade 9 and above

See Activity 14 or 16 for octahedron and Activity 18 for string design.

BIBLIOGRAPHY

Cundy, H. Martin, and A. P. Rollett. *Mathematical Models.* New York: Oxford University Press, 1961.

Laycock, Mary. *Dual Discovery through Straw Polyhedra.* Palo Alto, Calif.: Creative Publications, 1970.

Lyng, Merwin. *Dancing Curves.* Reston, Va.: National Council of Teachers of Mathematics, 1978.

National Council of Teachers of Mathematics. *An Agenda for Action: Recommendations for School Mathematics of the 1980s.* Reston, Va.: The Council, 1980.

Schmidt, Annette. *String Art: No Nails Needed.* Malvern, Pa.: Instructo/McGraw Hill, 1981.

Seymour, Dale, Linda Silvey, and Joyce Snider. *Line Designs.* Palo Alto, Calif.: Creative Publications, 1974.

Sharpton, Robert E. *Designing Pictures with String.* Buchanan, N.Y.: Emerson Books, 1974.

Stonerod, David. *Puzzles in Space.* Hayward, Calif.: Activity Resources Co., 1982.

String Art Encyclopedia. New York: Sterling Publishing Co., 1976.

Winter, John. *String Sculpture.* Palo Alto, Calif.: Creative Publications, 1972.